A PRACTICAL INTRODUCTION TO
THE DYEING AND FINISHING OF WOOL FABRICS

IAN BEARPARK, F WILLIAM MARRIOTT
and JAMES PARK

SOCIETY OF DYERS AND COLOURISTS

PO BOX 244, PERKIN HOUSE, 82 GRATTAN ROAD, BRADFORD BD1 2JB

1986

Typeset by Wessex Typesetters
(Division of The Eastern Press Ltd)
Frome, Somerset

Printed in Great Britain by
The Eastern Press Ltd
London and Reading

ISBN 0 901956 44 9

Preface

Although wool accounts for a relatively small proportion of the world's production and consumption of textile materials, and wool usage varies according to geographical location, as a fibre it is used to produce high-quality and high-specification materials. The dyeing and finishing of wool fabrics is therefore an important area of wool technology, adding much to the technical performance and the aesthetics of the final product. In addition many finishing processes and routines are specific to wool.

This monograph, which continues the series of practical publications on wet processing, is devoted to the dyeing and finishing of wool fabrics. In recent years many new products and processes have been developed, largely through the efforts of organisations such as the International Wool Secretariat, and perhaps this sector of the textile industry has in recent times made the biggest strides from science-assisted to science-based operations.

The authors would like to acknowledge the assistance of the members of the Chemical Processing Section of the IWS Development Centre at Ilkley, particularly Dr M A Rushworth, for assistance and encouragement in the preparation of this monograph. We are also grateful to the following companies for permission to use illustrations: Böwe Maschinenfabrik GmbH, Augsburg (Figure 4.7), L Ph Hemmer GmbH, Aachen (Figure 2.26), ILMA SpA, Schio (Figures 5.2 and 5.3), Neil & Spencer Ltd, Leatherhead (Figure 4.6), Holmes Heaton & Co. (Huddersfield) Ltd (Figure 2.7) and Rudolf Then GmbH, Schwäbisch Hall (Figure 5.4). Figures in Chapter 7 are taken from *An Introduction to Textile Finishing*, Second Edition, by J T Marsh, published by Chapman & Hall Ltd.

J Park
Series Coordinator

February 1986

THE AUTHORS:

Ian Bearpark, formerly finishing manager, West Riding Fabrics Ltd, Guiseley, Leeds

F William Marriott, senior development officer, International Wool Secretariat, Development Centre, Ilkley, West Yorkshire

James Park, managing director, Dymatecs Ltd, East Goscote, Leicester

Previous titles in this series published by the Society of Dyers and Colourists:

1. A Practical Introduction to Yarn Dyeing, by James Park
2. A Practical Introduction to Fibre and Tow Coloration, by Geoffrey Clarke
3. Dyehouse Laboratory Practice, by T M Thompson
4. An Introduction to Management in the Dyeing Industry, by James Park

Contents

CHAPTER 1

Introduction

Wool is an exceptionally important constituent of high-quality fabrics, either alone or in wool-rich blends. Statistics show that wool's share of total textile production (now about 5%) has slowly fallen [1]. In Western Europe wool accounts for about 12.5% of the total, i.e. just over 200 000 tonnes annually.

1.1 STAGES OF DYEING FOR WOOL

Dyeing of wool can be carried out on loose stock, top or sliver, yarn, fabric or garments. Dyeing in the fibre form for woollen-spun yarns and top or sliver for worsted spinning are undoubtedly the cheapest methods and are least critical in terms of level dyeing. However, with fibre dyeing lead time from choice of shade and dyeing to the delivery of finished fabric can be excessive, but it is the only stage at which the so-called heather-dyed effects can be introduced. Fibre dyeing has been extensively discussed in another monograph in this series [2].

Yarn dyeing is nearer to the delivery point and is suitable for the production of marl yarns. Dyeing in hank form is costly and productivity is limited by the slowness of associated winding, although these problems can be overcome by dyeing in package form. Adequate levelness must be obtained and it has been shown recently that package dyeing methods are capable of producing wool yarns that are satisfactory for a multitude of end uses [3]. Package dyeing methods also offer advantages for certain types of yarn, for example, those spun on open-end or other modified systems, where yarn strength may be slightly lower than that of conventional yarns. Yarn dyeing has been covered in a previous monograph in this series [4].

Dyeing of knitted and woven goods, either in fabric form or as made-up garments and hosiery, is important in the wool industry. The piece dyer is close to the point of sale and therefore can respond quickly to the dictates of fashion. Level dyeing is essential and the process costs are midway between those for fibre and yarn methods. Piece or garment dyeing can be integrated with finishing since nearly all wool materials, irrespective of spinning method, blend constituents, fabric-forming method or stage of dyeing, require some form of wet and/or dry finishing before sale.

Statistics collected by the International Wool Secretariat (IWS) indicate that wool consumption is divided between the various manufacturing sectors as shown in Table 1.1, which also indicates the percentage of the production in each sector that is dyed in fabric form. From Table 1.1 it can be calculated that 23% of all the wool produced is dyed by the methods described in this monograph.

TABLE 1.1

Product Distribution

Manufacturing sector	*Total wool consumption (%)*	*Proportion of sector production piece dyed (%)*
Carpets	24	
Woven woollen fabrics	21	42
Woven worsted fabrics	30	37
Knitted fabrics	3	14
Knitted garments	15	15
Hand-knitting yarns	7	

1.2 FASTNESS

The earlier the process stage at which dyeing is introduced, the less critical is the compromise necessary between level dyeing and attainable fastness. Increasingly stringent in-service requirements have made the selection of high-fastness dyes essential, even in the later process stages demanding high standards of level dyeing. Shrink-resist processes, applied to overcome the major defect of wool in washing, can be used at all the stages where dyeing takes place. Dye selection must be such that fastness is adequate if these processes are applied after dyeing and in all cases the wash fastness must be in line with the washing treatment which the shrink-resisted goods are designed to withstand physically.

1.3 OBJECTIVES OF WOOL FINISHING

The finisher's aim is to produce a fabric that the customer will find attractive in handle, appearance and weight. He must obtain the most desirable handle that the wool quality will allow, and in many cases the finished fabric will be altered to such a degree as to be unrecognisable from the original loom-state fabric. This change is generally more obvious in finished woollens than in worsteds, hence the common saying that 'woollens are made in the finishing, whereas worsteds are made on the loom'.

Finishing has undergone a quite drastic change during the last decade with the advent of modern high-speed, less labour-intensive machinery, which can be programmed for a high degree of reproducibility. This to a great extent has placed greater emphasis on the skill of the head finisher, and has been necessary because of the passing of the traditional textile finishing operatives whose skills were often handed down from father to son.

One important aim of this monograph is to outline the basic finishing processes in a straightforward manner as a guideline into this fascinating and often neglected area of the textile industry.

Although a monograph in this series is planned to cover the finishing of cotton fabrics, it is perhaps worthwhile underlining here the vast differences in finishing routines for cotton and wool fabrics. The former often follow fairly simple processing routes while the latter are typically subjected to complex finishing procedures, the end point of which can often only be judged subjectively.

REFERENCES

1. Ford, Dyer, **167** (March 1982) 24.
2. Clarke, 'A Practical Introduction to Fibre and Tow Coloration' (Bradford: SDC, 1982).
3. Wool Sci. Rev., No. 58 (1982).
4. Park, 'A Practical Introduction to Yarn Dyeing' (Bradford: SDC, 1981).

CHAPTER 2

Fabric Preparation

It has often been stated that, for successful dyeing, thorough preparation of woven or knitted fabric is essential. This is especially true for raw wool fabrics, from which lubricating oils and associated dirt must be removed by scouring, but other important preparatory processes are necessary, as described below. In this chapter, Sections 2.1 to 2.9 deal specifically with the preparation of woven fabrics.

2.1 PRE-FINISHING INSPECTION

The loom-state fabric should be carefully inspected as soon as possible after weaving so that any obvious faults can be noted and corrected on the loom. An assessment of the amount of knotting and mending is made, and then the fabric is accurately weighed and measured. The weight and dimensions of the fabric should be marked clearly on its face at the piece end, together with the piece and quality identification numbers, so that the amount of scouring and milling required for attaining the correct finished length and unit weight of the fabric can be calculated. The remaining unknown is the percentage weight loss which will occur during the finishing operations. This loss can vary from zero in the case of some dry-spun worsteds to about 20% in the processing of low-grade reprocessed woollen goods.

The finished length (pay length) can be calculated using Eqn 2.1:

$$F = \frac{W - L}{Y} \times 1\,000 \qquad (2.1)$$

where W = weight of loom state fabric, kg
L = expected loss in weight, kg
Y = required yield of finished fabric, g/m
F = length of finished fabric, m.

Pieces are usually assembled into like qualities for subsequent processing.

2.2 PROCESSING SEQUENCES

For woven fabrics the important preparation processes are scouring and milling; colour wovens and fabrics to be piece dyed require these processes. For certain constructions, such as velours, partial raising or other mechanical treatment may be given before dyeing. Setting processes may be undertaken to improve fabric stability during dyeing. Carbonising, if necessary, may be carried out at several stages in the processing sequence as shown in Figure 2.1. There are arguments for carbonising at various points in the manufacturing sequence, not least of these being the possible savings in chemicals that result from the use of carbonising acid residues for dyeing, the ease of impurity removal and the production of satisfactory dyeing quality.

	B	*C*
Carbonise Scour Alkaline mill Dye	Scour Carbonise Acid mill Neutralise Dye	Scour Carbonise Alkaline mill Dye
D	*E*	*F*
Scour Alkaline mill Carbonise Neutralise Dye	Scour Alkaline mill Dye Carbonise	Solvent scour Carbonise Neutralise Dye
	G	
	Scour Carbonise Semi-neutralise Acid mill Dye with level-dyeing acid dyes	

Figure 2.1 – Typical preparation sequences

Sequence A, referred to as greasy carbonising, allows for the maximum removal of cellulosic contaminants but requires that the lubricating oil is not removed by acid solutions. This restricts the choice of oil, in practice, to saponifiable types. However, by carbonising before scouring and alkaline milling, it does ensure that the wool is then at the correct pH for dyeing by all the major dye classes. Scouring before carbonising (sequence B) means that a larger range of types of lubricating oil can be used by the spinner, but since consolidation of the fabric may occur in the process, burr removal may be more difficult with highly contaminated materials. Acid and alkaline milling techniques (sequences B and C) give fabrics of different quality, acid milling being used when dense felting is required. When carbonising occurs just before dyeing as in sequence D or F, neutralising is necessary if dyes other than level-dyeing or 1:1 metal-complex acid dyes are being used. Sequence G gives significant economies in acid consumption.

Greasy carbonising has often been claimed to give least damage, owing to the protective action of the grease, but long-term storage of pieces in highly acid or alkaline conditions before dyeing should be avoided to prevent wool damage and yellowing. Acid treatment followed by an alkaline treatment, without due care, can cause serious wool damage. It has been suggested that greasy carbonising is best used in conjunction with subsequent solvent scouring. In practice solvent scouring is seldom applied.

A practical evaluation of the various processing routes listed in Figure 2.1 revealed that sequence D gave the best results, although all the sequences gave fabric of commercially acceptable quality [1]. This analysis was based on carbonising efficiency and dyed fabric quality, physical properties of finished fabric, fastness and handle. The sequence chosen in given circumstances may well depend on:

(a) Amount of burrs present and their ease of removal
(b) Storage time envisaged before neutralising or dyeing
(c) Types of dye or dyeing process to be employed
(d) Relative cost and saving in chemical consumption of any individual method.

2.3 CARBONISING

During growth wool may pick up various forms of foreign material, such as burrs and seeds; these can be removed from the fabric by treatment with sulphuric acid which chars the cellulosic impurities. The practical process consists essentially of the following stages:

(a) Impregnation of the material with acid
(b) Drying following by baking to char the cellulosic material
(c) Crushing to remove the cellulose as a powder – often carried out in milling machines
(d) Neutralisation of the residual acid

Modern practice is to use a continuous process for stages (a) and (b) with crushing being carried out discontinuously. Stage (d) is often incorporated in the scouring sequence. Machinery for continuous carbonising has been described [2] consisting of the following components:

1. Impregnation by means of squeeze rollers
2. A skying device to allow for thorough impregnation
3. A multi-stage tenter type of dryer to allow for drying, followed by baking at higher temperatures.

A modern range is shown in Figure 2.2.

Acid treatment should be carried out by impregnation with 7–8% sulphuric acid solution at room temperature for 5 min, followed by squeezing to 75–80% liquor retention, giving an acid uptake of 5% o.w.f. Typical running speeds are 5 m/min for heavyweight and 25 m/min for lightweight fabrics.

It has been indicated (see Figure 2.3) that immersion for 2–5 min at 35°C, in an acid solution containing 0.01–0.02% non-ionic wetting agent, provides the

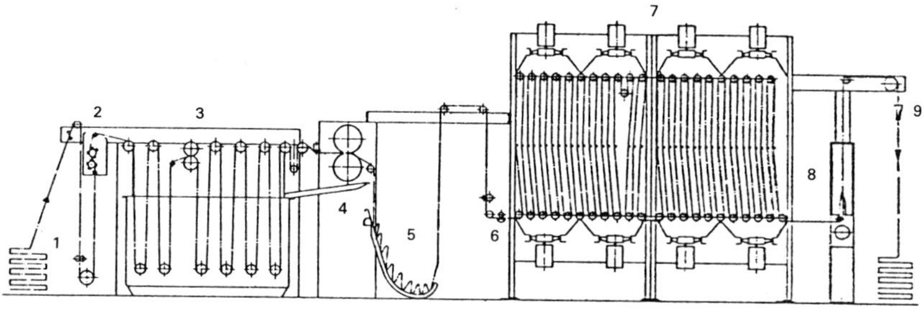

1 – Guider
2 – Set of cleaning brushes
3 – Impregnation tank
4 – High squeezing padder
5 – J-section for intermediate storage
6 Equipment to neutralise selvedges
7 – Drying carbonising chamber
8 – Cleaning chamber
9 – Exit pleater

Figure 2.2 – Continuous carbonising range

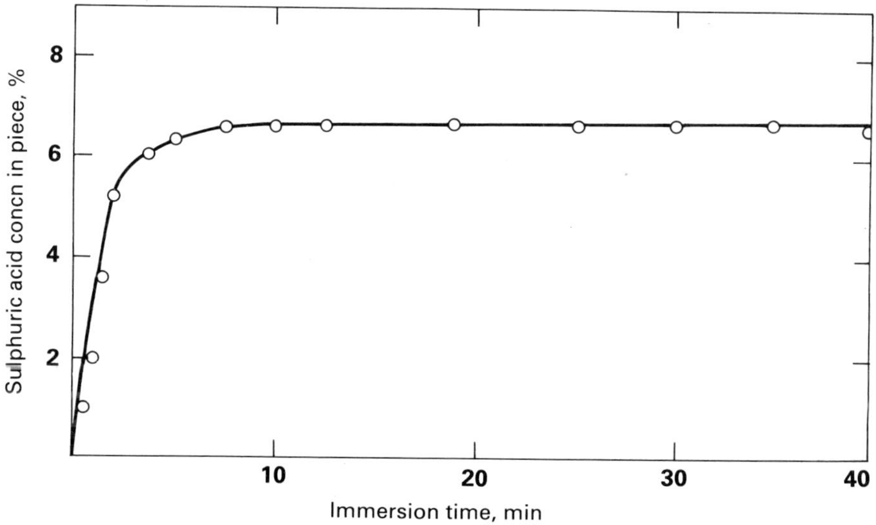

Figure 2.3 – Graph showing absorption of sulphuric acid (strength 4.4%) by woollen piece goods

optimum conditions for least damage of the wool and maximum absorption of acid by the burrs. The fabric should be dried to a 10% moisture content at temperatures up to 90°C and baked for 3 min at 130°C. Crushing is carried out for 60 min.

Much work has been carried out on the carbonising process [3]. The moisture content of the wool as it enters the dryer after acid impregnation can greatly influence the extent of wool damage. The significance of surfactant additions in the acid bath is less obvious. Satisfactory processing conditions have been defined, but anomalous results have been obtained by several investigators of the process.

2.4 SCOURING

Early surveys indicated that wide variations existed in practical scouring methods [4, 5]. Although the principal machine used has for many years been the dolly, variations exist in factors such as:

(a) Roller composition – materials include wood, cast iron, rubber, ebonite, cork, plastic, copper, fabric on cast iron or brass
(b) Roller diameter and nip pressure
(c) Fabric speed
(d) Machine capacity
(e) Time and temperature of scouring
(f) Time, temperature and volume of rinsing bath.

Scouring is often the first stage of a finishing sequence with the objective of producing a thoroughly clean fabric, free from processing oils and dirt; impurity levels should be less than 2% for woollen fabrics and 1% for worsteds. The dyed yarns in colour wovens will be much brighter in colour and the cover will be greater. If the degree of cover is insufficient, scouring should be prolonged and, if necessary, the fabric must be milled.

The chemistry of scouring wool fabric depends largely on the type of lubricant used in yarn spinning. When saponifiable oils rich in free fatty acid have been used, a saponification scour is employed in which the oil is converted to soap *in situ* by the addition of sodium carbonate (soda ash). The chemical reaction between the oil and the alkali produces the soap, together with the formation of carbon dioxide. The type of residue produced is often claimed to enhance the handle of the fabric. With non-saponifiable oils, including those of the self-emulsifiable type, scouring depends on the combined action of a surfactant and an alkali to remove the oil from the fabric and to keep the oil and associated dirt particles in suspension.

2.4.1 Scouring Method

Carbonised pieces are usually run in cold water for 15–30 min followed by draining to remove the excess acid. This will normally reduce the acid content of the fabric to about 3%. Generally saponification scouring is carried out cold for 20 min with the addition of soda ash only. Emulsion scours with alkali and detergent are carried out at 40°C. Hot and cold rinses complete the process. Colour woven fabrics are scoured cold to avoid the possibility of colour bleeding. After the scouring stage the oil and dirt in suspension in the detergent liquor must not be redeposited on the fabric since their removal becomes difficult. Rinsing is carried out by the slow dilution of the scouring liquor.

Stains are usually removed by spotting with a solvent-based agent prior to scouring to aid their removal. For heavily stained or dirty pieces a solvent-based detergent is incorporated into the scouring liquor. Moxon has described the preferred routines for the various scouring techniques [6] and these are still valid in the main, although more use is now made of synthetic detergents and self-emulsifiable oils.

2.4.2 Scouring Calculations

Scouring is generally carried out at a liquor ratio of about 2:1, with soda ash used when saponifiable oils are present. The amount of alkali used is calculated on the basis of the amount required to accomplish the following four functions:

- To neutralise free residual acid from carbonising or dyeing, *a*
- To neutralise acidic side chains in wool keratin, *b*
- To act as a detergent builder, *c*
- To saponify the oils present in the wool, *d*.

Total required = $(a + b + c + d)$ kg soda ash

$$b + c \simeq 5 \text{ g/kg wool and}$$

$$d = \frac{FC}{50} \text{ g/kg}$$

where F = free fatty acid content of the oil, %
C = oil content of the wool, %.

For non-saponifiable oils, including self-emulsifiable types, soda ash must be added to satisfy factors *a*, *b* and *c*, but scouring is carried out with the addition of 1–2% detergent.

The use of excessive alkali should be avoided since this can lead to harshness of handle and yellowing during drying or subsequent storage. However, sufficient alkali must be present to give stable emulsions during scouring so that the dirt associated with the fabric is removed with the lubricant. Unremoved or redeposited dirt is difficult to remove later without the oil with which it is usually in close contact.

2.4.3 Practical Scouring

Woollen Scouring

Greasy woollen fabrics contain 3–9% oleine oil which is applied at the fibre blending stage to provide lubrication during carding and spinning. The oil quality is important to the scourer, for lower-grade spinning oils generally contain a higher percentage of mineral oil which can be troublesome, especially if used for whites that are to be piece dyed. Mineral oils may become redeposited unevenly on the fabric giving rise to streaky dyeings. The scourer should impress upon the spinner the importance of using good-quality oleine as this can reduce substantially reprocessing problems during dyeing and finishing.

A general scouring procedure should be introduced by the scourer for all fabrics, and this should be adjusted according to the amount of cover required. Various soaps and detergents can be used, depending on the finish required. The most commonly used scour is the two-bath scour. The following typical procedure refers to the scouring of a fabric containing 9% oleine on a dolly washer running at about 100 m/min.

(a) The fabric is loaded into the scouring machine using cross-drafting (see Figure 2.4).

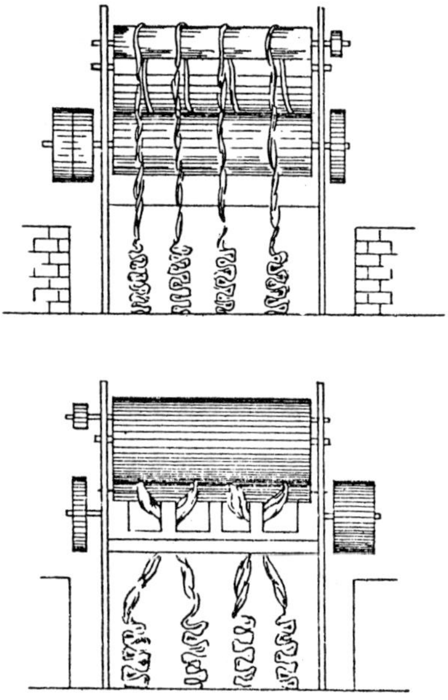

Figure 2.4 – Cross-drafting

(b) A solution of 20–25 g/l soda ash (sp. gr. 1.020–1.025) is added to the vessel using 2 l/kg fabric, a liquor ratio which is ideal for scouring. The fabric should be scoured in this liquor for 30 min, with the liquor retained in the scouring vessel.

(c) Water at 45–50°C is then run into the vessel, with the top suds drain open to permit the dirty liquor to run to waste. Warm washing-off should be carried out for 30 min followed by 10 min cold washing-off.

(d) Soda ash solution (sp. gr. 1.020–1.025) at 1.5 l/kg fabric is then added to the vessel with additional non-ionic detergent or soap. The fabric should be run in this liquor for 25–30 min with the liquor retained.

(e) Washing-off in warm water, with the top suds drain open, is then carried out for at least 35 min, followed by a cold water wash-off for 10–15 min, the fabric being removed from the vessel when cold. When scouring whites for piece dyeing it is advisable to add a sequestering agent after the first 25 min of warm washing. It is good practice to acidify the washing bath after 5 min cold washing and to continue running in the acidified liquor for the final 10 min. The goods are acidified for the following reasons:

1. White fabric is less likely to yellow during tentering
2. Pile fabrics for raising are in an ideal state for this process
3. Grey mixtures remain clear and are less likely to be affected by 'Crombie yellowing' (i.e. bleeding of coloured on to white fibres in the blend)
4. There is less likelihood of bacterial attack
5. Plaids or tartans are less likely to cross-stain if left for any time in a wet state before tentering.

It is not advisable to acidify the rinsing bath before piece dyeing with neutral-dyeing and milling acid dyes as they may strike too quickly and give uneven dyeings. These five steps are typical of woollen scouring in a dolly washer.

Worsted Scouring

Little or no spinning oil is applied to worsted yarns nowadays. Such fabrics seldom contain more than 1% extractable matter and soda ash treatment is not essential. In practice, however, the scourer will use soda ash, but in much reduced quantities, for it has good detergency properties and acts as a soap builder. The following is a typical procedure.

(a) The fabric is wetted out with warm water and soda ash solution (sp. gr. 1.020–1.025) added at a rate of 0.5 l/kg fabric. A suitable soap or soap-detergent should be added to increase the scouring action and to assist fabric cover development. Scouring is carried out for 10 min in this liquor.
(b) Washing-off is carried out at 45–50°C for 10 min.
(c) More surfactant is added and scouring continued for 15 min.
(d) The fabric is washed-off in warm water for 15 min followed by a cold water wash-off for 5 min, during which the pH should be adjusted to 5.5–6.0 with the addition of acetic acid. When cold, the fabric should be removed and then hydroextracted/mangled and tentered as soon as possible.

Some important basic rules should be adhered to so that scouring faults can be kept to a minimum.

(a) Softened water should be used at all times to prevent the formation of lime soaps which will result in uneven piece dyeings, tacky handle and an unpleasant smell in the finished fabric.
(b) Sufficient liquor should be present in the washer at all times during

scouring. This will enable the fabric to balloon, thus preventing the formation of running marks or creases.

(c) Heavier fabrics require more nip pressure than lightweight qualities. An increase in pressure can assist development of fabric cover, but if excessive pressure is applied there is a greater risk of forming running marks.

(d) It is good practice to cross-draft pieces during scouring, as this helps to prevent running marks and at the same time increases the amount of cover.

(e) Pieces should be removed from the washer only when they are cold. If pieces are taken from the washer in the warm state and stored for any length of time then persistent creases can result, especially with dark heavy fabrics.

(f) When loading the fabrics, it is good practice to insert one twist into each rope. More than one twist can cause the fabric to roll into a tighter rope, resulting in crease formation and unevenness in subsequent piece dyeing. Rolling can also take place when the fabric has a tight listing, due to faults in poor warping or weaving practice.

2.4.4 Fabric Conditions

For satisfactory further processing, the following limits are normally applied to scoured pieces:

- Oil content, $\leqslant$0.5%
- Soap content, 0.75–1.0%
- pH of pieces for dyeing, 6.5–8.0
- pH of pieces for raising prior to dyeing, 4.0–5.0.

2.4.5 Scouring Machines

Although there has been interest in solvent treatments and continuous techniques [7], batchwise aqueous scouring is still the principal method. In principle, scouring is carried out by passing the saturated fabric through a nip at high speed. The water is forced from the fabric carrying with it soap and suspended dirt, which is collected in a trough below the nip and then deposited to the drain. The action of the water bursting from within the fabric, together with the nip pressure, results in loosening and felting of the fibres, producing fabric cover.

This basic operation is carried out in various types of machines. Conventionally the dolly washer was used, as illustrated in Figure 2.5. In the rope scouring machine illustrated, A and B are squeeze rollers, C is the main trough and D is the internal or suds trough in which it is possible to reduce the effective liquor ratio for scouring by preventing the return of scouring liquor to the main trough. E and F are guide rollers and G is the draft board with an automatic knock-off mechanism if pieces become tangled; H shows the fabric path. A number of ropes are scoured simultaneously. In rope scouring

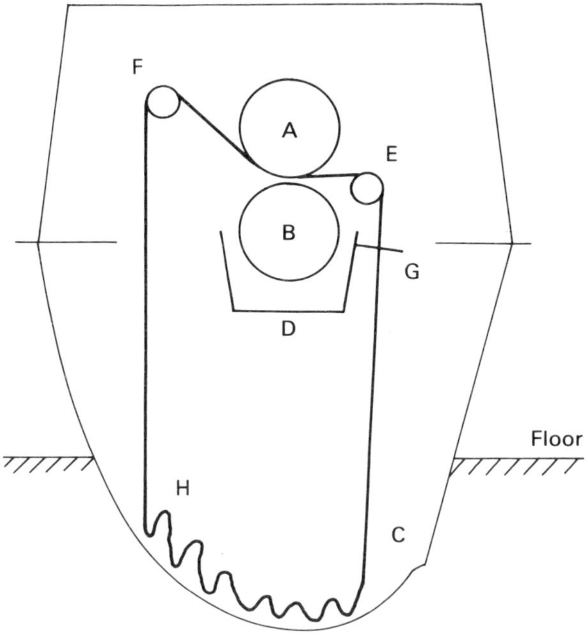

Figure 2.5 – Scouring machine

methods fabrics that are susceptible to selvedge curling and the formation of running marks are bagged for scouring, which involves sewing the selvedges together along the length of the piece with the face of the fabric inwards.

Rollers A and B can be made from the same or different materials and numerous possibilities exist. Considerable variations in roller diameter and weight of top rollers are encountered. The fabric speed also varies but is generally around 100 m/min. Traditional becks were made from jointed wooden staves. Although stainless steel or resin coatings have been introduced more recently, it has been argued that the wooden becks produce fabrics with more cover and superior handle.

Open-width scouring machines can also be used and, although the principle of the machine is similar, they tend to be used mainly for fabrics which will develop creases or running marks in rope scouring. Such a machine is illustrated in Figure 2.6.

The latest high-speed scouring machine can be run at three different speeds, a low speed (100 m/min) for wetting out, a medium speed (150 m/min) for lighter scouring and a high speed (200 m/min) for normal scouring. As the fabric is circulating at least twice as fast as in a dolly washer, the scouring time can be cut by 50% or more. The top roller pressure can be varied to suit the type of fabric by means of a pneumatic arrangement which gives the machine great versatility, so that most types of woollens and worsteds can be processed on one machine.

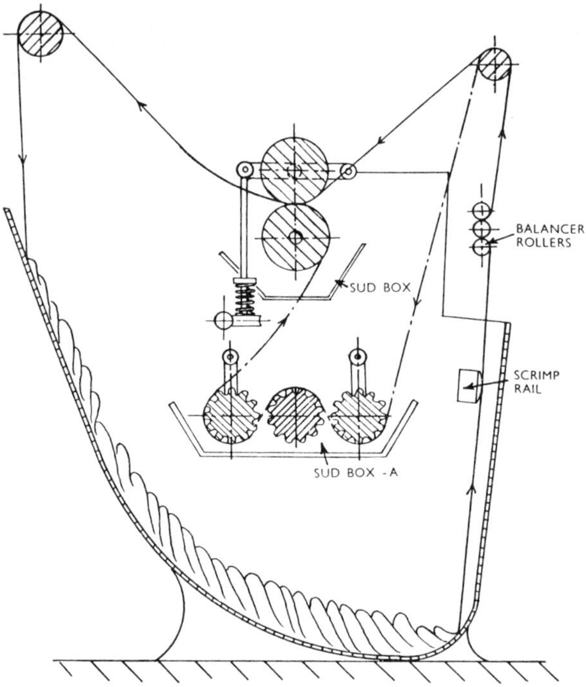

Figure 2.6 – Open-width scouring machine

One of the great advantages of these modern machines is the availability of a card-operated program which enables the finisher to achieve excellent reproducibility. Programs can be created for various styles and this enables the soda ash, soaps or detergents, hot and cold water, and drains for suds box and main tank to be automatically added and controlled to give trouble-free scouring. The operative has only to load and unload the vessel, which means that relatively unskilled labour can carry out satisfactory scouring with very little training.

A baffle plate can be lowered into the path of the fabric at the exit from the roller nip to produce a semi-milled effect on the fabric. This baffle can be lowered or lifted at any time during the scouring program so that varying amounts of cover can be achieved.

Dispensing tanks can be fitted to the high-speed scouring machine. These enable the scourer to prepare a scouring liquor whilst the machine is being loaded, and great economies can result by recycling and storing the warm water from the final rinsing operation. This liquor can quite safely be used during the first scour of the following load, resulting in savings in steam, water and effluent that can be quite considerable.

High-speed scouring machines are available from a number of

Figure 2.7 – Combined scouring and milling machine

manufacturers. Nowadays these machines are being superseded by combined scouring and milling machines of the type illustrated in Figure 2.7 (see Section 2.5.6).

2.5 MILLING

Milling or felting is a property associated with the characteristic scale structure of the wool fibre. The differential frictional effect means that each fibre is able to move more readily in the direction of its root during the milling process. The effect of milling can vary from a slight surface felting for further development in raising processes to the production of dense felts. Milling is carried out under alkaline (pH 10) or acid (pH 2) conditions in the presence or absence of a detergent. Acid milling gives a dense felt and each process type produces its own individual handle. The soaps formed as a result of saponification scouring can improve the effectiveness of the subsequent milling process.

The objects of the milling process in wool fabric manufacture are:

(a) To improve fabric strength; a loom-state fabric is normally weaker than a milled fabric made from it

(b) To enhance surface appearance; this can be dramatic as in the so-called blind finish

(c) To increase the density of the fabric in terms of its weight per unit length; this is particularly evident in the production of billiard fabrics which may show 15% warp and 35% weft shrinkage

(d) To improve handle (i.e. softness) by producing a cover; this is sometimes merely a consolidation of the yarn rather than a thorough felting
(e) As a necessary preliminary to raising.

There are three important practical milling processes: grease, soap and acid milling.

2.5.1 Grease Milling

Grease milling is a cheap but dirty method of carrying out the milling operation. The greasy fabric is run through a liquor containing 25–30 g/l soda ash (sp. gr. 1.025–1.030) and then squeezed to a liquor ratio of 1:1 or 3:2. The fabric is then milled to the required cover and dimensions before washing. In this process the fabric is milled in the presence of the dirt and loose colour normally washed out in the scouring process. As this can result in dirty pieces, especially in grey mixtures and block checks, it is seldom used for medium- or good-quality fabrics.

2.5.2 Soap Milling

Most finishers consider that the softest and most attractive fabric handle is obtained by soap milling. Sodium (or potassium) soaps are dissolved to give a 5% solution in hot water. Such solutions have to be applied warm as the soap solidifies on cooling. Care must be taken if the fabric has been carbonised, to ensure that thorough neutralisation has taken place, for acidified fabric will flocculate the fatty acids in these conditions. The soap acts as a lubricant and decreases the inter-fibre friction, permitting more rapid shrinkage than in grease milling with alkali alone. Soap is considered essential for producing heavily consolidated fabrics such as meltons and billiards. During lengthy soap milling, however, certain fabric constructions cease to shrink further. In these circumstances it is advisable to wash-off the fabric in a washer and then to return it to the milling machine. Excessive lubrication by soap can 'block' the felting process by reducing the influence of the directional frictional effect. A typical processing sequence is summarised in Table 2.1.

TABLE 2.1

Processing Sequence for Soap Milling

Stage	*pH*
Scouring	10
Carbonising	2
Neutralising	7
Soap milling	10
Washing-off	7
Piece dyeing	3

Newer blended milling soaps have become more popular in recent times because of the ease of application, less blotching and ease of removal after milling. Conventional washing-off times after milling have been halved by using these blended products, and economies in energy and effluent costs can be made.

2.5.3 Acid Milling

Acid milling is generally preferred for woollen fabrics for the reasons discussed below. A typical processing sequence is summarised in Table 2.2.

TABLE 2.2

Processing Sequence for Acid Milling

Stage	*pH*
Scouring	10
Carbonising	2
Washing-off	3
Acid milling	3
Piece dyeing	3

Whenever the fabric pH is changed from alkaline to acid, a potential processing problem exists; comparing Tables 2.1 and 2.2 shows that this is twice as likely to occur in soap milling than in acid milling. Acid is preferred as a milling medium because:

(a) The resultant fabric has a greater tensile strength
(b) The process occurs more rapidly and to a greater extent
(c) Milled fabrics show a greater degree of extensibility and stretch
(d) The wool fibres suffer less degradation, giving less flocking and therefore heavier fabrics than with soap or grease milling
(e) At low pH there is less chance of dyes bleeding or cross-staining
(f) Washing-off can be minimised, resulting in lower water consumption
(g) Power costs are lower because less weight is necessary on machine lids and rollers
(h) Pieces remaining wet for a time are less likely to result in bacterial damage or colour bleeding
(i) Grey mixtures are less likely to yellow.

2.5.4 Factors Outside the Finisher's Control

These can be summarised as follows:

(a) Wool fibre length and fineness (the finer the wool, the more readily it will felt)
(b) Yarn tenacity, extensibility and elastic recovery

(c) Yarn density and structure; yarns that are tightly twisted are more resistant to felting
(d) Fabric density and structure, i.e. density of warp and weft yarns, and type of weave; for instance, plain weave construction will mill less readily than a 4/4 twill.

Such factors are outside the control of the finisher who should discuss these implications with the yarn and fabric producers if the desired finish specification proves difficult to achieve.

2.5.5 Factors Within the Finisher's Control

The following factors need to be controlled by the finisher.

Temperature
Heat is necessary to promote rapid milling, which is carried out at 40–50°C, but water must be added during the process to compensate for losses by evaporation. The temperature of the fabric during milling can be controlled by opening or closing the machine door and, in some instances, by means of steam pipes located within the vessel.

pH and Milling Agents
Wool felts least in neutral or slightly acid conditions. The rate of felting is slow within the pH range 4–7 and reaches a minimum near pH 4.8, the isoelectric point of wool. Under these conditions wool is best able to withstand the rigours of the milling operation.

The fabric should be in the pH ranges 2–3 for acid milling, 9–10 for grease or soap milling to ensure good felting properties. If felting should occur too quickly or unevenly the process can be slowed down by milling at pH 6–7 so that a uniform finish is obtained.

Liquor Ratio
The liquor ratio recommended for satisfactory milling will vary according to fabric construction and machine design, but this variation is usually within the range 1:1 to 2:1. More liquor will retard felting and increase the danger of skidding and running marks. Insufficient liquor results in excessive flocking and the risk of mechanical damage and mill rigging.

Fabric Dimensions
The miller can control the rate and direction of the felting operation by altering machine settings as follows. To increase width shrinkage and reduce length shrinkage the milling machine lid is raised, the roller pressure increased and the mouthpiece closed. It is also possible to tie up the roller before the mouthpiece so that a certain amount of 'drag' or 'brake' is asserted on the

warp, thus pulling out any length milling. The fabric rope is run in multiple drafts.

On the other hand, the length shrinkage can be increased and the width shrinkage reduced by lowering and weighting the milling machine lid to increase the load on the fabric passing through the spout. The mouthpiece is opened as wide as possible and the machine run with just enough load on the rollers for satisfactory running performance. The fabric is run up in a single draft.

It is good practice for the miller to bring the width and length measurements down together and it should always be the aim to run the fabric for the last 10 min or so with the milling machine lid down as this will assist in the prevention of running marks. Uniform dense 'blind' felting is obtained by slow and gradual shrinkage. If felting occurs too rapidly, effective 'bottom' milling is not achieved.

Fabric Handling and Measurement of Dimensions

Machine loading and fabric lay out in a rope milling machine can have a considerable bearing on the final finish obtained by the miller. It has been mentioned that the fabric can be run singly or in multiple drafts, even up to four drafts with lightweight fabrics. Care must be taken to level the pieces so that the fabric will not knot up and cause damage. When fabrics are to be milled for long periods they should be reversed evey hour or so as this will help in preventing running marks.

Certain constructions induce the fabric to roll tightly in rope form and in such cases the fabric should be bagged (stitched selvedge to selvedge), care being taken to ensure that this is done efficiently. The sewing should be squared off, for long-sided pieces will give rise to mill rigs and ripped selvedges during milling. The bagging procedure will result in ballooning of the tubular 'bag'; if this becomes excessive the fabric may be torn. The remedy for this fault is to leave unsewn gaps of about 0.5 m along the edges which will allow the trapped air to escape.

Pieces are usually sewn together or end to end using an industrial sewing machine as illustrated in Figure 2.8.

Most modern milling machines have fairly accurate measuring clocks fitted to the front of the machine. Where this is not the case measuring strings are tied into the selvedge to indicate changes in fabric length; strings initially 1 m apart simplify the estimation of percentage changes. Width shrinkage is easier to determine as the fabric can be laid out on the floor and accurately measured by rule.

The miller must also consider the processes that the fabric must undergo after milling. Fabrics may continue to lose width and gain length during subsequent piece dyeing or raising. The prolonged washing necessary after soap milling can have adverse effects on the fabric dimensions. Only after careful monitoring of the first two or three milling trials with an unfamiliar fabric construction can the finisher collate sufficient information to ensure that the fabric can be milled routinely to yield the desired finished weight and

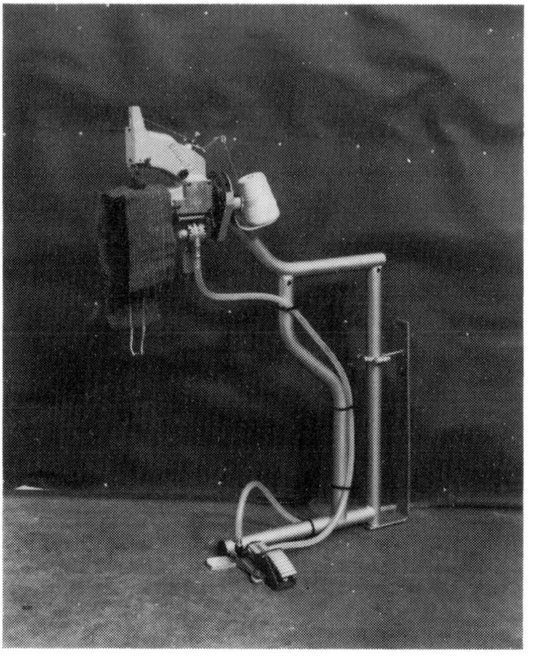

Figure 2.8 – Industrial sewing machine

dimensions after further processing. Finished fabric data should always be available to the miller so that a check can be kept on:

(a) Target finished length (pay length, m)
(b) Increase in length during finishing (m)
(c) Target finished width (cm)
(d) Expected loss in weight (%)
(e) Target finished yield (g/m).

This information will enable the miller to constantly adapt and slightly alter his milling parameters should the need arise.

2.5.6 Milling Machines

Faller Stocks

Milling is carried out in batchwise aqueous processes, and the earliest machines developed for this purpose were the stocks, illustrated in Figure 2.9. These are extremely useful for the production of dense felts by acid milling methods. In the faller stocks the fabric is placed in the semicircular machine and the heavy wooden feet pound the fabric and in doing so rotate it slowly. The control of width and length shrinkage is more difficult than with rotary milling, but the amount of weight loss incurred is considerably less because

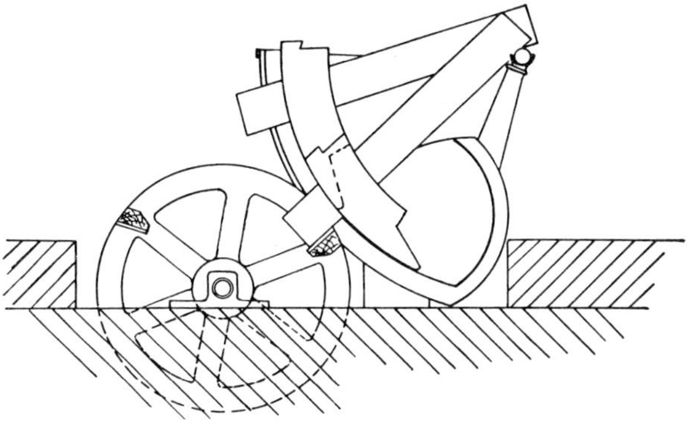

Figure 2.9 – Faller stocks

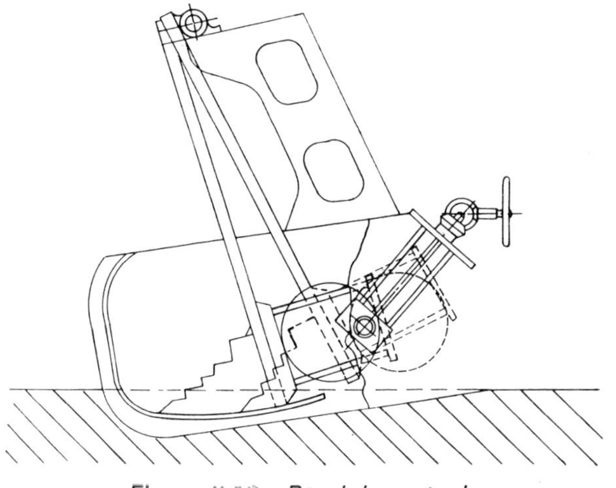

Figure 2.10 – Pendulum stocks

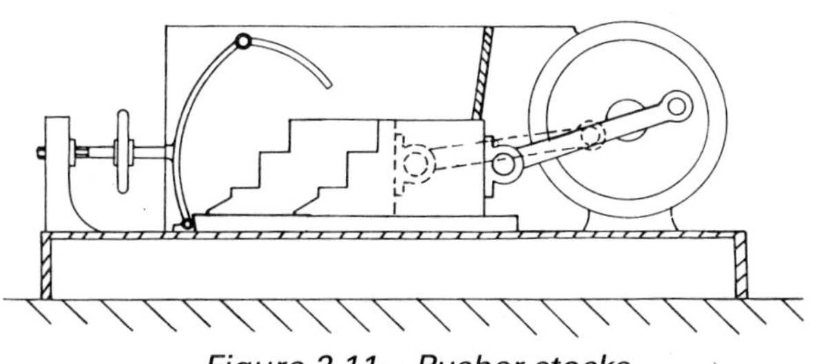

Figure 2.11 – Pusher stocks

the pounding causes less flocking than does running in rope form. A fine soft handle is obtained which is ideal for wool/angora blends and for jersey fabrics, and whilst there is no danger of running marks care must be taken to avoid 'crow's foot' marking. The fabric is loaded side to side for width milling and end to end for length milling. Stocks can be used to correct faulty fabrics (run marked or mill rigged) by running end to end for 20 min followed by a hot open-width scour.

More sophisticated versions of the stocks are the pendulum stocks shown in Figure 2.10, and the pusher stocks shown in Figure 2.11, which have a positive driven action.

Rotary Milling

In the rotary milling machine, illustrated in Figure 2.12, the milling action depends on the generation of heat during the process, the control of pH and

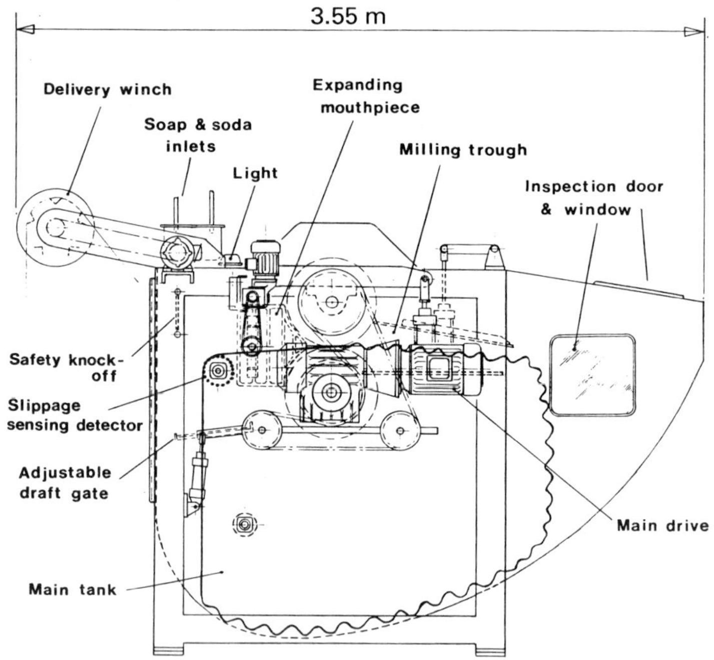

Figure 2.12 – Rotary milling machine

liquor ratio, and the careful application of pressure. An increase in roller pressure causes weft shrinkage, whereas greater pressure in the spout gives warpway shrinkage. The initial felting must not be too rapid or crease marks may be formed.

The most commonly used machines for milling woollen fabrics are the single-roller and the tandem-head machines. The fabric is loaded, usually double or triple draft, and run until the correct length and width measurements have been achieved, by careful adjustment of position and loading of the milling lid and roller jaws.

Rotary milling performance varies from machine to machine and with various machine characteristics:

(a) Wooden or plastic rollers
(b) Flanged bottom roller to give better and quicker width milling
(c) Single rollers or tandem rollers; tandem rollers can more than double the rate of felting and are ideal for heavy milling on billiards or meltons (see Figure 2.13).

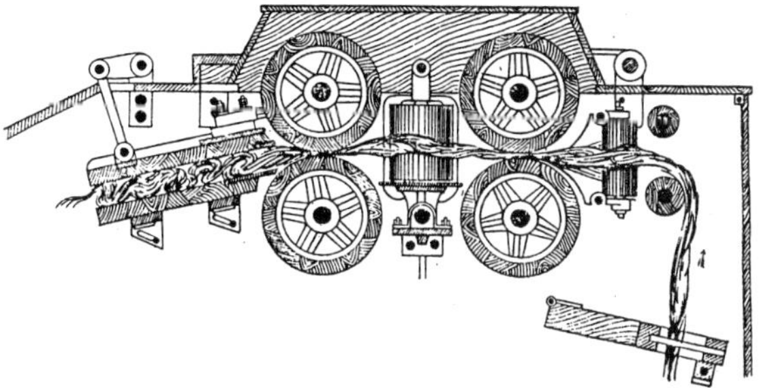

Figure 2.13 – Tandem roller machine

To avoid milling rigs the correct settings and running procedures should be adopted, and constant supervision should take place. Maintenance on the rotary milling machine is critical as incorrect settings can result in severe damage to the fabric.

Combined Scouring/Milling Machines

These machines are rapidly taking the place of the conventional rotary milling machines in the newer finishing mills. As the name suggests, the fabrics, often six, eight or twelve pieces at a time, are loaded in the machine and then scoured and milled in one operation. This is possible because the rollers are used not only for milling, but with increased pressures can be used to squeeze the pieces after scouring and prior to milling.

These machines are highly productive and are very economical to run because of the savings made in water consumption. The liquors can be recycled or stored to use for the following batch. A machine of this type is illustrated in Figure 2.14.

Combined machines, whilst offering savings and higher productivity to the finisher, demand greater responsibility from the machine operator, since he is processing six or more pieces per load. To enable him to obtain satisfactory milling performance from each piece in a load it is important to process together fabrics of closely similar quality and length. Any variations will present the operator with problems, and it is likely that milling variations will result.

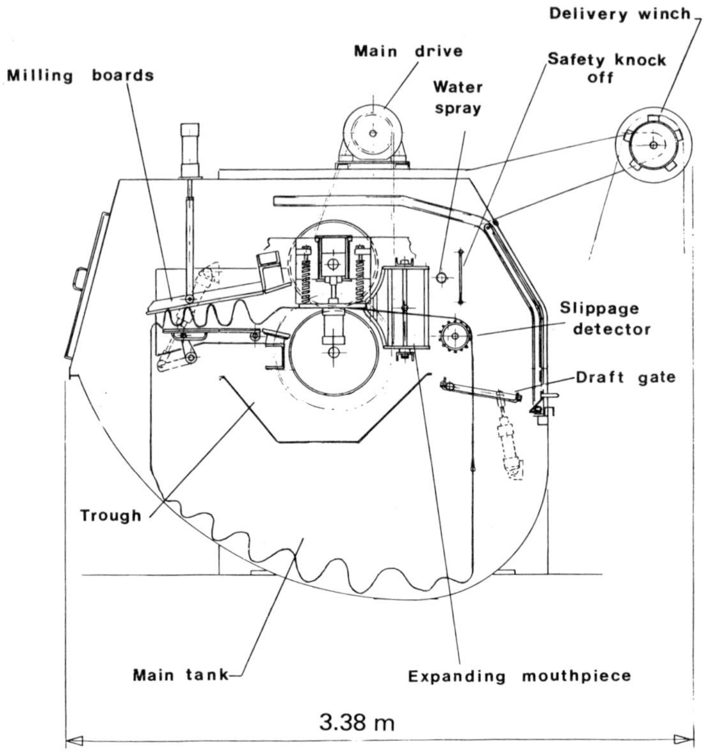

Figure 2.14 – Combined scouring and milling machine

Combining scouring with milling simplifies the entire finishing sequence, as follows:

(a) Greasy carbonise and bake
(b) Neutralise–scour–mill–wash-off in one operation in the combined scour/milling machine
(c) Tenter.

Taking the whole range of wool fabric qualities, the selection of an appropriate milling machine for a given fabric is very much 'horses for courses'. The choice varies from one finishing plant to another and the entire milling operation still remains 'A wonderous mystique of the skillful millerman', as it always will be.

Wira has carried out much work on piece scouring [4–6] and milling [8]. This showed that, provided the wool has good felting properties, satisfactory milling can always be achieved within a reasonable time if the conditions of treatment and machine settings are appropriate. Milling machines with flanged rollers are advantageous for certain fabrics. Variations in yarn structure, weave, scouring treatment and washing-off do not affect significantly the character of the milled fabric.

2.6 SETTING PROCESSES

2.6.1 Crabbing

When some clear-finish worsteds are scoured the tensions imparted in the fabric production processes are quickly relaxed. This results in cockling and 'crow's footing', i.e. width distortion of the design and weave. To avoid such faults, which cannot be easily corrected, the fabric should be crabbed before scouring. During the crabbing operation the fabric is wet set so that the pre-scoured appearance is maintained, and this particularly applies to the successful finishing of fine-weave fabrics such as mohairs, tropicals and clear-finished gabardines.

In this process the fabric is permanently set by treatment in roll form with boiling water and steam (see Figures 2.15 and 2.16). The machine often

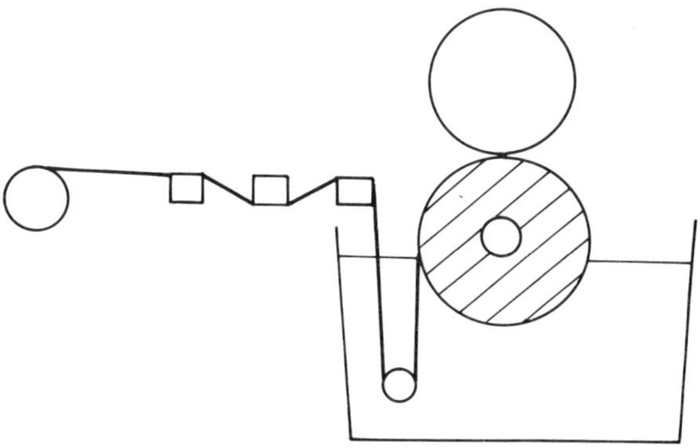

Figure 2.15 – The Yorkshire crab

consists of two crabbing units, two steaming units and a cold water quench. The fabric, to which cotton end pieces have been attached, passes under even tension in open width by way of a guide roller through the boiling water bath and is wound on the underside of a roller about 60 cm in diameter, so that the rotating fabric roll is half immersed in the water. The fabric then passes to a second crabbing unit to ensure uniformity of treatment. It is then wound on a perforated steaming cylinder covered with cotton fabric, and steaming is carried out for 10–15 min with steam at 0.1–0.5 MPa (1–5 atm) pressure, with the fabric rotating throughout the process. To ensure uniformity the fabric is run to a second steaming cylinder and the process repeated. Finally it is quenched in cold water. Crabbing and steam blowing must be carried out evenly if level dyeing is to be obtained.

1 – Straining beam
2 – Cloth guider
3 – Hot water trough
4 – Water trap
5 – Expander roller
6 – Steam case
7 – Guide roller
8 – Endless rubber blanket
9 – Fabric expander device
10 – Cooling trough
11 – Squeezing rollers
12 – Guide roller
13 – Transport hasp
14 – Plaiter
15 – Heating drum
15a – Condensate drain
16 – Cleaning device for endless blanket
17 – Cleaning device for heating drum
18 – Regulating device for endless blanket

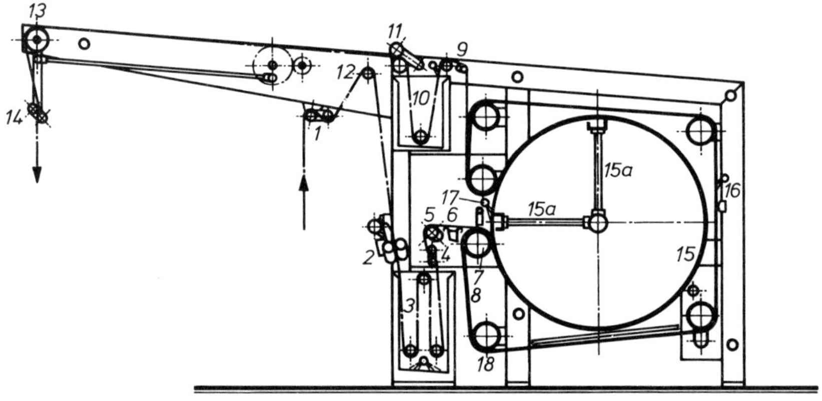

Figure 2.16 – The Konticrab (Hemmer)

2.6.2 Potting

This process is somewhat similar to crabbing and is used for high-quality woollen fabrics. The fabric is wound carefully onto a cotton-wrapped perforated iron roller and the roll placed in a tank of cold water. The temperature of the water is then raised slowly, to either 75–80°C or the boil, and treatment continued for 3–5 h. After cooling and rewinding to reverse the fabric, the operation is repeated. Rust from the iron rollers can give staining problems.

2.7 DRYING BEFORE STORAGE

Since many pieces are stored for prolonged periods before dyeing and finishing, they are dried at this stage for storage. This is usually carried out by rotary hydroextraction or squeezing followed by tentering (see Section 6.1.2). It has been found that a 'hot mangling' can give increased drying efficiencies since the tenters can be run at higher speeds. The temperature of the hot water is 70°C for ecru fabrics but should not exceed 45°C for colour woven and piece-dyed fabrics.

It has also been found that tentering of pieces between finishing processes minimises the formation of creases and running marks.

2.8 RAISING BEFORE DYEING

For certain qualities such as velours the final appearance of the fabric is enhanced if some finishing such as raising is carried out prior to dyeing. Wet or dry raising may be employed, and this process is discussed later (see Section 7.2). For such qualities the processing sequence may be: scour, carbonise, mill, dye, wet raise, tenter, dry raise, cut and press; alternatively a wet raise may be carried out before dyeing.

2.9 FABRIC STORAGE

Fabrics are often stored for prolonged periods when manufacturing instructions are received before shade selection is finalised. They are thus partially processed and stored to reduce the delivery time once the shades have been selected. Lemin [9] examined the influence of storage conditions on the risk of subsequent dyeing faults. It was found that pieces should be covered to avoid damp and draughts, they should have an adequate regain, and a pH of 4.5 is preferred. Water spotting on wool containing residual acid can cause hydrolytic degradation.

2.10 KNITTED FABRICS

Knitted fabrics are normally produced from cleaner yarns and contain small amounts of emulsifiable oils. Scouring is usually carried out in the dolly at 40°C with 0.5% soap or synthetic detergent. Light milling is then given to obtain the necessary degree of surface cover.

2.11 BLEACHING

Wool fabrics are bleached by batchwise methods on the winch by running for 4 h at 50°C in 4.5 vol. hydrogen peroxide stabilised to pH 8–9 with sodium pyrophosphate. Continuous processes may be used in which the fabric is padded through a hydrogen peroxide bath containing acid, followed by drying and storage in the dark for 48 h.

REFERENCES

1. Park, J.S.D.C., **87** (1971) 114.
2. Sedgwick, Dyer, **126** (1961) 731.
3. Park, J.S.D.C., **87** (1971) 111.
4. Moxon and Barritt, 'A Survey of Scouring and Milling in the Wool Industry', Wira Publication (June 1950).

5. Moxon, 'The Scouring of Woollen and Worsted Piece Goods', Wira Publication 223 (July 1964).
6. Moxon, Wira Bulletin, No. 22 (1960) 2, 63.
7. Wool Sci. Rev., No. 23 (1963) 13.
8. Moxon, 'An Account of Wira's Work on Milling', Wira Publication 216 (March 1961).
9. Lemin, J.S.D.C., **86** (1970) 169.

CHAPTER 3

Dyeing Processes

The choice of dye class and dyeing method for wool fabrics depends on the following factors:

(a) Preparatory processing, whether carbonising or shrink-resist treatment has been given
(b) Fabric density, the degree of milling will affect penetration
(c) Fastness to further processing and in-service requirements
(d) Design and efficiency of available equipment.

The general principles of dye selection have been discussed elsewhere [1]. The main classes of dyes for wool are listed in Table 3.1, with comments on their properties.

TABLE 3.1

Dye Classes for Wool

Dye class	*Application*
1:1 Metal-complex	Require a low pH; this is advantageous for dyeing uniformly carbonised wool without pre-neutralisation, good coverage of neps; good fastness
Level-dyeing acid	Can be applied at low pH; good penetration; suitable for carbonised goods; low wet fastness but high light fastness and bright colours possible
Milling acid, 1:2 metal-complex and reactive	High fastness; suitable for shrink-resist end uses; carbonised fabric must be neutralised; careful control of pH and temperature required during dyeing and level-dyeing assistants usually employed
Chrome	Economically attractive; high fastness in full depths; possible environmental problems with effluent

Adequate wetting out of the fabric before dyeing is essential. With carbonised fabrics a pre-rinse may be given to remove excess acid. Uniform neutralisation may be difficult to achieve, so that appropriate choice of dyes and process is important. The level-dyeing properties of 1:1 metal-complex and level-dyeing acid dyes at low pH can be used to advantage on uniformly carbonised fabrics which have not been neutralised. With most classes of dye, however, it is necessary to run the fabric to even out the pH before proceeding. Least damage to the wool is incurred if dyes are applied in the isoelectric region of pH 4.2–4.8.

3.1 1:1 METAL-COMPLEX DYES

The Neolan (CGY), Palatine Fast (BASF) and Inochrome (ICI) dyes are 1:1 metal-complex azo acid dyes and are applied from a sulphuric acid bath at pH 2. The use of higher pH values gives tippy dyeings of inferior fastness. The conventional application technique is to wet out the fabric at 60°C for 15 min in a bath containing 8% o.w.f. sulphuric acid. The dye is added and the temperature is raised to the boil over 30 min and dyeing continued for a further 90 min. The material is then thoroughly rinsed. These dyes are particularly suitable for carbonised goods when dyeing follows immediately after carbonising and the acid from that process can be used as part of that required for dyeing.

Modified application methods have been introduced to minimise the risk of acid damage while maintaining the level-dyeing and fastness properties. Dyeing can be carried out using 4% sulphuric acid [2] either with Albegal NF (CGY) at the boil or with Irgasol HTW (CGY) to protect the wool at 110°C.

Reasonably bright trichromatic combinations can be selected from the wide range available, but these dyes can cause difficulties with metamerism. A typical tertiary combination is C.I. Acid Yellow 54, Red 183 and Blue 158.

3.2 LEVEL-DYEING ACID DYES

Acid dyes are generally classified into four application groups depending on the optimum dyebath pH (see Table 3.2). The pH for optimum level dyeing rises as the neutral-dyeing affinity of the dye is increased.

TABLE 3.2

Application Conditions for Acid Dyes for Wool

Dye group	*Bath addition*	*Final bath pH*
1	Sulphuric acid	2.5–3.0
2	Formic acid	~4.0
3	Acetic acid	4.5–5.5
4	Ammonium sulphate	6.0–8.5

The level-dyeing acid dyes in group 1 are applied from a bath containing 3–5% sulphuric acid and 10–20% sodium sulphate. Dyeing is begun at 60°C, the bath is raised to the boil in 30 min and boiling continued for 45 min. Group 2 dyes are applied by a similar technique except that 2–5% formic acid is used. Group 3 and 4 dyes belong to the milling acid type as discussed later.

TABLE 3.3

Selection of Levelling Acid Dyes for Wool

Good coverage of carbonising faults C.I. Acid	*Good levelling but poor fastness C.I. Acid*	*Higher fastness and reasonable levelling C.I. Acid*	*Special higher fastness dyes C.I. Acid*
Orange 3	Yellow 17	Yellow 29*	Yellow 25*
Red 80	Yellow 23	Orange 53	Red 130
Blue 40	Red 1	Red 57*	Red 145
	Red 37	Red 73	Violet 4
	Red 52	Red 240	Blue 271*
	Violet 7	Red 289	Black 26:1
	Blue 2	Violet 41	*Also*:
	Blue 7	Blue 67	Erio Fast Rubine 2GE* (CGY)
	Blue 52	Blue 72*	Xylene Brilliant Green BGE (S)
	Green 16		Supramin Navy Blue BN (BAY)

* Tertiary combinations

Level-dyeing acid dyes can be selected (see Table 3.3) to give maximum coverage of carbonising faults, maximum levelness or sufficient fastness to meet the minimum Woolmark specification [3]. In general a three-component mixture of primary dyes is used for matching.

3.3 MILLING ACID DYES

Milling acid dyes are applied from a dyebath at pH 5.5–6.0 obtained by the use of acetic acid and a levelling agent, such as 2–3% Albegal SW (CGY). Dyeing is begun at 50°C, the bath raised to the boil in 50–60 min and dyeing continued for a further 30 min. One possible rationalised selection of dyes from this range is given in Table 3.4.

TABLE 3.4

Milling Acid and 1:2 Metal-complex Dyes for Wool

Milling acid dyes C.I. Acid	*1:2 metal-complex dyes C.I. Acid*
Yellow 79	Yellow 129
Red 111	Orange 86
Red 134	Red 219
Violet 48	Green 74
Blue 80	Brown 19
Blue 138	Black 58
Green 25	

3.4 1:2 METAL-COMPLEX DYES

Neutral-dyeing 1:2 metal-complex dyes were introduced in the 1950s; they give high fastness properties without the application and disposal problems associated with chrome dyes. The 1:2 metal-complex dyes can be applied at pH 5.5–6 with acetic acid or at pH 6–8 with ammonium sulphate together with a suitable levelling agent. One possible rationalised selection of conventional 1:2 metal-complex dyes is given in Table 3.4.

Recent developments in this dye class have seen the introduction of dyes with anionic solubilising groups. The Elbelan N (HDC) dyes are symmetrical 1:2 complexes solubilised with two carboxylic groups, and the Acidol M (BASF) dyes are symmetrical complexes solubilised with two sulphonate groups. The Lanacron S (CGY) dyes and Neutrichrome S (ICI) dyes are assymmetrical 1:2 complexes solubilised by one sulphonate group. The advantages of all these ranges of dyes include high tinctorial value, high wet fastness in full depths and economy in manufacture.

The washing fastness of 1:2 metal-complex dyes can be further increased by aftertreatment in a fresh bath for 20 min at 50°C with 10% Sandopur SW (S). The availability of anionic 1:2 metal-complex dyes aftertreated in this way has extended the range of dyes suitable for machine-washable wool. The development of metal-complex dyes has been reviewed recently [4].

3.5 USE AND MODE OF ACTION OF LEVELLING AGENTS

Level dyeing of wool is achieved by careful control of dyebath pH and temperature. Levelling agents are usually also necessary for dyes of high fastness but poor levelling properties. Such products assist in overcoming small-scale non-uniformity (skitteriness) associated with the differential uptake of dye at the root and tip of individual fibres, as well as larger scale unlevelness such as uneven dyeing across the width of the fabric (listing). Levelling agents permit dyes to be applied at slightly lower pH values than normally used for a particular dye class. This can be useful when applying neutral-dyeing dyes to carbonised wool. It also enables dyes to be applied in the isoelectric region of pH 4.2–4.8 where minimum damage to the wool fibre occurs.

There are two main types of levelling agent:

(a) Anionic agents with substantivity for wool; these compete with the dye anions for the available sites in the fibre and are gradually displaced as dyeing proceeds

(b) Cationic or non-ionic agents capable of complexing with anionic dyes; as the temperature rises during dyeing, free dye anions are slowly released from the complex and taken up by the fibre.

The second type of product is probably the easiest to control and leaves no residues on the fibre that might cause yellowing or loss in fastness.

3.6 CHROME DYES

Before the development of metal-complex dyes, chrome dyes were widely used to produce colours that showed good fastness to washing, particularly in heavy depths. Three application methods were available, but only the afterchrome process is still significant:

(a) Chrome mordant method, where the wool is treated with the chromium compound and the dye subsequently applied
(b) Metachrome method, in which the dye and the chrome mordant are applied simultaneously
(c) Afterchrome method, in which the chrome dye is applied in a similar manner to an acid dye and then treated in a chrome bath.

The disadvantages of chrome dyeing are as follows:

(a) Lengthy processes, sometimes requiring fresh baths
(b) Tendering of the wool on prolonged dyeing
(c) Severe changes in colour during chroming
(d) Difficulties in correction of faulty dyeings
(e) Health hazards due to the handling of chromium compounds
(f) Difficulties in disposal of effluents.

Chrome dyes have been extensively discussed by Bird [5]. The use of these dyes is now restricted to the production of economical maroon, brown, navy colours and black by the afterchrome method with minimum application of dichromate [6]. A typical process of this kind involves setting the dyebath at 40°C with:

- Acetic acid (80%), 3%
- Formic acid (85%) (to give a pH of 3.5), 1%
- Alcosist PL (Allied Colloids), 0.5%

The material is circulated for 10 min, the dye added, the temperature raised to the boil over 40 min and circulation continued for a further 20 min. If necessary, for complete exhaustion, a further 1% formic acid (85%) is added and dyeing continued for 10 min longer. The bath is cooled to 75°C and the

TABLE 3.5

Some Chrome Dyes for Wool

Shade area	*C.I. Mordant*
Maroon	Red 7
Navy	Blue 47
Brown	Brown 27
Black	Black 9

with shading dyes such as C.I. Mordant Yellow 5

calculated minimum amount of potassium dichromate is added. The pH is adjusted to 3.5–3.8 with formic acid (85%) and the temperature is raised to the boil over 15 min. After 10 min at the boil 7.5% sodium sulphate is added and chroming continued at the boil for 15 min. The bath is then cooled and a cold rinse completes the process. Suitable dyes are listed in Table 3.5.

3.7 REACTIVE DYES

Although fibre-reactive dyes were originally developed for cellulosic fibres, several commercial ranges of reactive dyes have been developed specifically for wool (see Table 3.6). Reactive dyes display a tendency to skittery dyeing on

TABLE 3.6

Reactive Dye Ranges for Wool

Dye range	*Year launched*	*Reactive group*	*Auxiliary*
Lanasol (CGY)	1966	α-Bromoacrylamide	Albegal B
Drimalan F (S)	1969	2,4-Difluoro-5-chloropyrimidine	Lyogen FN
Verofix (BAY)		*N*-Methyltaurine-ethylsulphone	Avolan REN
Hostalan (HOE)	1971	β-Sulphatoethylsulphone	Eganol GES

wool and a number of auxiliary products have been developed to assist level dyeing or to improve dye uptake. The development of these dyes has recently been reviewed [7]. The Lanasol range has probably been the most widely used and a typical application method is as follows. The dyebath is prepared at 40°C with:

- Ammonium sulphate, 4%
- Albegal B (CGY) as levelling agent, 2%
- Albegal FFD (CGY) (de-aerating agent), 0.5%
- Acetic acid, if necessary
- Glauber's salt, 10% when using less than 1% dye, 1% for 1–2% dye, nil for more than 2% dye.

The material is circulated for 10 min, the pH checked and adjusted to 6.5 for pale depths, or to 4.5 for heavy depths. The dye is added, circulation continued for a further 5 min and the temperature raised to 70°C over 30 min. Dyeing is continued at this temperature for 20 min, then the bath is raised to the boil over 30 min and dyeing continued for a further 60 min. When shade matching is satisfactory the dyebath is cooled to 80°C, the pH raised to 8–8.5 by adding 2% ammonia and the fabric treated for 20 min to remove unreacted dye. The process is completed by hot and cold rinses. A suitable selection of dyes is given in Table 3.7.

Reactive dyes give bright colours of high fastness to light and wet treatment. They are widely used for dyeing shrink-resist-treated material.

TABLE 3.7

Lanasol Reactive Dyes for Wool

Lanasol dye	*C.I. Reactive*
Yellow 4G	Yellow 39
Orange R	Orange 68
Red 6G	Red 84
Red 2G	Red 116
Red 5B	Red 66
Blue 3R	Blue 50
Blue 3G	Blue 69

3.8 DYEING OF SHRINK-RESIST-TREATED FABRICS

The production of shrink-resistant materials is discussed in Chapter 9. Dyes for such fabrics must have sufficient fastness to correspond with the machine washability of the treated fabric and, if applied before the shrink-resist process, to withstand the conditions of this treatment. At the start of dyeing the fabric should be run in cold water and the pH checked. The pH is adjusted to 6.5 before starting dyeing, and in some cases no acid addition is required.

Reactive dyes have been widely used for bright colours, but the costs of dyes and processing are relatively high. Selected 1:2 metal-complex dyes, with or without the Sandopur SW treatment, and chrome dyes may be used for the production of deep, fast and economical colours.

3.9 IDENTIFICATION OF DYEING FAULTS

Problems arising during dyeing can result in the production of faulty fabric. In addition the dyeing process will show latent faults in the fabric that have been caused by previous processing. Faults must be classified so that corrective action can be taken and processes improved when possible. Examples of common dyeing faults include:

(a) Streaky or patchy dyeing arising from uneven scouring, milling or carbonising
(b) Variation in colour between pieces within a dyed batch because they have undergone different conditions of preparation
(c) Differences between repeat batches owing to weighing errors or variations in dyeing process conditions
(d) Unlevel dyeing because of incorrect dye selection or application method
(e) Off-shade dyeings caused by reprocessing to overcome unevenness.

3.10 CORRECTION OF DYEING FAULTS

Many of the above faults can only be corrected by reprocessing the fabrics, and in some cases the problems are difficult to rectify. Correct choice of dyes, auxiliaries and methods of preparation and dyeing will minimise the faults. Correction of unlevel, patchy or off-shade pieces usually depends on the application of a levelling procedure chosen according to the type of dyes used.

With level-dyeing dyes significant levelling can be obtained by boiling the fabrics in a blank dyebath. With the faster dyes, such as milling acid dyes, boiling in a blank dyebath with the addition of levelling agents may also prove effective.

Reductive stripping with 5% formic acid and 5% sodium sulphoxylate formaldehyde in a boiling bath may be necessary for fast dyes. This treatment may give non-uniform stripping and possible damage of the wool, so that the problem still remains after subsequent redyeing. With exceptionally fast dyes, such as reactives, the best course is to dye the fabrics to either a deeper colour or black.

Such levelling procedures are seldom useful when unlevel dyeing is a result of differential fabric treatment or damage in previous processes. Furthermore the stripping treatment may cause additional changes in dyeability, and the process may seriously effect the strength and physical appearance of the fabric. Rescouring before redyeing may give improvements, whilst a light milling before redyeing will improve the strength and physical appearance of the fabric.

3.11 REPRODUCIBILITY/FASTNESS/LEVELNESS

In any dyeing process it is the aim to achieve a level and on-shade material without the need for further colour additions or reprocessing. Much has been published with regard to the concept of using so-called blind dyeing techniques. These have been widely used in some areas of dyeing technology, but they are more difficult to apply in fabric dyeing, especially with wool, due to the number of preparatory processes which may be carried out prior to dyeing. The success of dyeing largely depends on the use of a standard and reproducible preparation sequence.

In general, however, reproducibility of dyeing is easier to achieve with the neutral-dyeing dyes, which possess excellent fastness but only moderate to poor levelling properties. Thus a high degree of reproducibility in wool dyeing depends on adequate control of preparation processes, choice of appropriate dyeing technique and selection of dyes to compromise between levelness, fastness and reproducibility.

3.12 DYEING OF BLEND FABRICS

A number of blended fabrics are dyed in piece form, the more important are discussed below.

3.12.1 Wool/Cellulosic Blends

The design is often such that, when the wool component is dyed using acid dyes only, a coloured and white check effect is obtained.

3.12.2 Wool/Nylon Blends

The nylon is included in many cases to improve fabric strength. Solid colours are obtained by the use of a blocking agent.

3.12.3 Wool/Polyester Blends

These are dyed with acid or metal-complex dyes for the wool and disperse dyes for the polyester. Alternatively pre-mixed dyes such as the Forosyn (S) range can be used.

3.12.4 Wool/Acrylic Blends

These are dyed with the usual dye classes for the wool and basic dyes for the acrylic. A useful method has been described [1] where Lanasol reactive dyes are used for the wool and the Maxilon M (CGY) migrating basic dyes are used for the acrylic portion.

The dyeing of blends has been extensively reviewed [8] and the methods for dyeing the above blends are similar to those described for yarn [1].

REFERENCES

1. Park, 'A Practical Introduction to Yarn Dyeing' (Bradford: SDC, 1981).
2. Soiron, Textilveredlung, **14** (May 1979) 189.
3. Park, Dyer, **147** (1973) 27.
4. Lewis, Wool Sci. Rev., No. 54 (Sept 1977) 30.
5. Bird, 'The Theory and Practice of Wool Dyeing', 4th Edn (Bradford: SDC, 1972).
6. Bayer Technical Information Note 479 (Feb 1977); Ciba–Geigy Technical Information Note (Nov 1977).
7. Lewis, J.S.D.C., **98** (1982) 165.
8. Shore, 'The Dyeing of Synthetic-polymer and Acetate Fibres', Ed. Nunn (Bradford: Dyers' Company Publications Trust, 1979) 411.

CHAPTER 4

The Processing of Hose and Garments

The comments regarding services and instrumentation made in Chapter 5 are applicable to the specialised operations of hose and garment processing. Two main types of woollen knitgoods will be discussed:

(a) Half-hose, which includes men's socks, children's socks, anklets, sports hose and children's three-quarter hose

(b) Garment blanks such as pullovers, cardigans, jumpers and their accessories.

The hosiery industry is divided into two sections, producing seamless or fashioned goods using circular or flat-bed machines respectively. One of the main problems of the garment dyer is that of penetrating seams and linkings. There is also the tendency for knitted fabrics to roll at the selvedges and this causes the seams to roll on one another.

4.1 DYEING AND FINISHING OF HALF-HOSE

After knitting, linking and mending, socks are received for dyeing with the 'right' side outwards. Coarse qualities and those containing cellulosic or synthetic-polymer fibres in the blend will normally be processed in this manner. Socks knitted from finer and softer yarn qualities are reversed to bring the 'wrong' side outwards. Turning proceeds at a rate of 400–500 pairs per hour by hand or 700–750 pairs per hour by machine. The socks are counted during the turning process and tagged with coloured thread or some other means to differentiate sizes.

Scouring used to be carried out in a tom-tom machine, but few of these now remain in use. Companies that still possess these machines use them for scouring woollen-spun socks, and for garments which may contain up to 10% oleine and require treatment with soda ash to saponify the oil to produce soap *in situ* as described in Chapter 2. Because of the gentle action of these machines they are also used in the scouring and milling of high-quality goods.

Scouring is normally carried out in either side-paddle or rotating-drum machines with 1% non-ionic detergent and either 1–2% ammonia or 0.5–2% soda ash at 50°C for 20 min or 5 min respectively. Times may vary according to quality and the amount of consolidation required.

Almost all wool socks are given a shrink-resist treatment (see Chapter 9). This may be applied either before or after dyeing, each having its advantages. Treatment before dyeing avoids colour changes due to the process or ancillary treatments with chemicals such as sodium bisulphite. Better handle and wearing properties are obtained if the socks are shrink-resist treated after dyeing, but the dyes used must be fast to the treatment. Whichever method is adopted, dyes of high wet fastness must be selected since most shrink-resist treatments cause modification to the wool fibre, so that desorption occurs more readily. Shrink-resist agents can be applied in conventional side-paddle or overhead-paddle dyeing machinery at room temperature.

Before the start of dyeing the socks are run in water only and the pH checked. The pH is adjusted according to the dye class being used. Milling acid, 1:2 metal-complex, chrome and reactive dyes are applicable (see Chapter 3 for methods). Where nylon reinforcing yarns are present, a blocking agent is added to the dyebath to minimise preferential dye uptake by the nylon. For light to medium shades 3–0.5% Erional NW (CGY) can be used. After dyeing at the top temperature for 20 min, the penetration of the heels and toes is checked and, if this is satisfactory, dyebath exhaustion is completed by the addition of acid. The dyebath is then cooled and the socks given a cold rinse. The socks, particularly coarser qualities, are given a treatment with a cationic softening agent, 0.5–1% being applied in a fresh bath at 50°C for 15 min. Permanent softeners are often used. Excess water is removed from the socks by rotary hydroextraction and then drying is carried out in a tumble dryer.

The socks are finally 'trimmed', in which process they are sorted into sizes and turned. The socks are put on wooden formers of the required size and are pressed between two heated metal surfaces at about 115°C for 1 min. The trimmer operates two sets of boards, stripping and filling one set of twelve whilst the second set is in the press. A trimmer can produce 100–150 pairs per hour, and in the industry this occupation has in the past been jealously guarded. The size of the former used must conform to the true size of the sock or relaxation shrinkage will occur in subsequent washing. Excess pressure during pressing produces glazing, whilst insufficient pressure leads to inadequate setting with subsequent curling of heels and toes.

4.2 DYEING AND FINISHING OF GARMENTS

Cardigans, pullovers and sweaters are frequently made up as garment blanks in the grey (ecru, undyed) state or containing dyed yarn if the blanks are for finishing only, with necks and stoles knitted separately for sewing on after finishing. The various sizes need to be carefully identified with tags and the accessories kept with the appropriate blanks. Generally necks are linked in dozens and the garment blanks turned 'wrong' side outwards for processing.

The accessories (necks, stoles, etc.) are usually placed in bags for dyeing, and occasionally the garments themselves may be placed in bags where a minimum of marking on the face is required, or where distortion is known to be a problem. About one dozen garments are placed in a bag.

The processing routes vary according to type of wool, in-service specification and stage at which the wool is shrink-resist treated.

4.2.1 Wool Type

Three types of wool are typically used for garments: botany (worsted-spun merino), lambswool (either woollen or worsted spun) and Shetland.

Botany Garments
These are usually of the machine-washable category, the processing sequence being:
(a) Anti-cockle
(b) Scour
(c) Chlorinate (if wool not shrink-resist treated)
(d) Dye (if required)
(e) Resinate (where wool has not been previously shrink-resist treated, and the highest shrink-resistance category is required, e.g. IWS Superwash category)
(f) Soften
(g) Finish out.

Lambswool Garments
The usual processing sequence is:
(a) Scour/mill
(b) Chlorinate (for machine-washable garments)
(c) Dye (if required)
(d) Resinate (if required for the highest level of shrink-resistance)
(e) Soften
(f) Finish out.

Shetland Garments
Dyeing is normally carried out on the yarn or loose stock. The usual processing sequence is:
(a) Scour/mill
(b) Shrink-resist treat
(c) Soften
(d) Finish out.

4.2.2 In-service Specification

There are three levels of specification:
(a) Dry clean only
(b) Hand wash
(c) Machine washable.

For dry clean only and hand wash garments some measure of dimensional stability is necessary. A shrink-resist treatment is essential for machine-washable garments (see Chapter 9 for details).

4.2.3 Stage at which the Wool is Shrink-resist Treated

Obviously if the wool has been given a shrink-resist treatment already as loose stock or top a further application is unnecessary.

4.2.4 Anti-cockle Treatments

This treatment is necessary for certain types of knitted structure which will distort during wet processing. Such garments are typically treated for 10 min at the boil in a side-paddle machine at a liquor ratio of 30:1 with 1–3% sodium metabisulphite and 3% non-ionic detergent. The steam is then turned off and the garments allowed to stand in the bath for 10 min. The liquor is circulated again, the bath is cooled to 60°C for 10 min and the garments rinsed and drained.

4.2.5 Scouring

Scouring procedures and equipment similar to those already described for hosiery are suitable. Alternatively the following procedure can be used:
- Detergent, 2 g/l
- Tetrasodium pyrophosphate, 2 g/l
- Sodium sulphate, 1 g/l

for 6 min at 40°C followed by a cold rinse.

4.2.6 Milling

Milling is carried out in several different types of machine, a comparison of processing times being shown in Table 4.1. A typical procedure uses 4 g/l soap

TABLE 4.1

Milling Times in Commercial Machines

Machine type	*Milling time (min)*
Rotating-drum	2–5
Pegg K-type	10–20
Tom-tom	10–30
Side-paddle	20–40

and 1 g/l tetrasodium pyrophosphate, run at 40°C to obtain the desired cover, followed by two rinses.

One-bath scour/mill recipes are often used where the oil content prior to treatment is not high or where it is known that self-emulsifiable lubricants

have been used. Proprietary scour/mill formulations are commonly used, which contain the following components:
(a) Detergent
(b) Water-softening agent
(c) Solubiliser
(d) pH buffers.

Two separate treatments at 40°C with 5 g/l scour/mill agent are given, followed by rinsing. Table 4.1 gives suitable milling times.

4.2.7 Shrink-resist Treatment and Dyeing

The comments already made for hosiery apply equally to garments. Reactive dyes are used for garments only where the highest fastness is required in full depths.

4.2.8 Drying

Excess water is removed from the garments by rotary hydroextraction and drying completed in a Kamsin-type tumble dryer at 70–75°C for 5–15 min.

4.2.9 Finishing

The finishing of garments is carried out in a manner similar to that described for hose, and is also known as trimming. Wool garments are usually processed on a flat-bed press operating at a steam pressure of 0.15–0.2 MPa (1.5–2.0 atm). The garments are first 'framed' to size and shape (ladieswear and menswear). The frames, preheated to avoid condensation, are constructed from stainless steel with the arms being hinged at the shoulder for ease of manipulation. The garments are loaded with the seams to the edges of the frame, steamed on the open steam table for 3–5 s and then cooled. The garment is then removed from the frame, laid out on the steam table and pressing is carried out at 102–105°C with the following sequence: steam for 5 s, cool, steam for a further 2 s if necessary, cool and blow for 3 s, and cool.

4.3 SOLVENT TECHNIQUES FOR GARMENTS

Methods of scouring, milling and shrink-resist finishing (see Chapter 9) from a solvent medium have been developed using machinery similar to that used for industrial dry cleaning. Machine loading for wool garments is usually only 60% of conventional loading, and even lower loading is used for delicate fabrics. As in dry cleaning processes, attention must be paid to the recovery and reuse of solvents.

4.3.1 Scouring

This is simply carried out by immersing the goods in dry cleaning solvent at room temperature in the absence of water or processing aids. Agitation is for 2–3 min, after which the soiled liquid is removed by extraction.

4.3.2 Milling

Milling takes place in the so-called charged systems, where water is present in the solvent. Recently three techniques have been described [1]: steam, solvent and low-surfactant milling.

Steam Milling

The goods are entered into the machine and steam is injected while the cage is rotated. The goods are milled for a predetermined period, before pure solvent is pumped into the machine and the goods are given a conventional scour. The goods are hydroextracted and dried in the normal way.

Conventional Solvent Milling

A milling bath is prepared by adding a milling aid, e.g. 10 g/l Avivan WA (CGY), to a known amount of solvent. The goods are scoured in clean solvent and extracted at high speed for 3–4 min. The milling liquid is introduced from the make-up tank. Water plus 1% milling aid (on weight of water) and the combined weight of the previous two components of perchloroethylene are mixed and placed in the additive box. The emulsified perchloroethylene and water are added to the tumbling goods and tumbling continued for a predetermined time, sufficient to give the desired level of cover. The goods are hydroextracted at high speed for 3–4 min, the milling solution is returned to the tank and the goods are dried.

Low-surfactant Milling

This method uses much less milling aid (5% on weight of water) than the method previously described. The goods are scoured in clean solvent and extracted at high speed for 3–4 min. Clean perchloroethylene is introduced from the main tank. A milling aid solution is prepared consisting of the required amount of water and 5% Alcosolve C (Allied Colloids) on the weight of the water, a little perchloroethylene is added to this and the mix well stirred. The emulsified perchloroethylene and water are added to the tumbling goods through the additive box and the goods tumbled for a predetermined time to give the required cover. The goods are extracted at high speed for 3–4 min, the milling bath being dropped to the still for recovery of solvent. The goods are finally dried.

Process Comparison

The steam milling technique is cheap since it requires no additional chemicals and gives even penetration with reduced processing times. However, water separation must be efficient and it is not suitable when using the same machine for resin applications where residual water affects the shrink-resist effect.

The conventional solvent milling method has the advantage that it is not necessary to distil the milling bath after every batch, but the milling bath can be contaminated by dark colours. Large quantities of milling aid are required, a tank is needed to store the milling liquid and the residual milling aid on the goods requires that a rinse be given before resin treatment.

With low-surfactant milling one solution must be redistilled after each batch. Only small amounts of milling aid are required, there is no contamination between batches and, due to the low concentrations used, there is no need to rinse the goods before the resin treatment.

4.4 MACHINERY FOR PROCESSING HOSE AND GARMENTS

Machines for processing hosiery and garments have much in common, as described below. Piece dyeing machines (see Chapter 5) are quite different in design and operation.

Figure 4.1 – Tom-tom scouring and milling machine

4.4.1 Scouring and Milling Machines

The oldest scouring and milling machine in use for hose and garments is probably the tom-tom (see Figure 4.1), which gives the material its own particular degree of consolidation. The goods are placed in the rectangular tub, usually constructed of wood, which moves to and fro whilst the cam-operated beaters fall under gravity.

A number of scouring machines are based on the design of industrial washing machines in which the garments circulate within a rotating drum. These machines usually include the facility to hydroextract, thereby reducing labour costs by eliminating the need to transfer goods between machines. Such machines permit a high degree of automation. They seldom have a heating facility; hot water must be supplied either directly or by means of a

Figure 4.2 – Cherry Tree Borderer

cold water feed and a steam mixing valve. Rotating-cage machines are regarded as having a severe action. Machines of this type include the Cherry Tree Borderer (see Figure 4.2), the Neil and Spencer Scourex and machinery from Thomas Broadbent.

4.4.2 Dyeing Machines

The most widely used machine for scouring, milling and dyeing of garments is the side paddle (see Figure 4.3). This consists of an oval stainless steel vessel

Figure 4.3 – Side-paddle machine

with a central island and a large paddle along one side of the machine to agitate the liquor. All chemical and dye additions to the machine are made by means of an addition hopper situated on the central island. Heat is supplied by open or closed steam coils and the agitation is usually variable. This machine is widely used for garments which have previously received a shrink-resist treatment at an earlier stage of processing. The overhead paddle (see Figure 4.4) is used for many of the simpler functions of the side paddle; it provides a

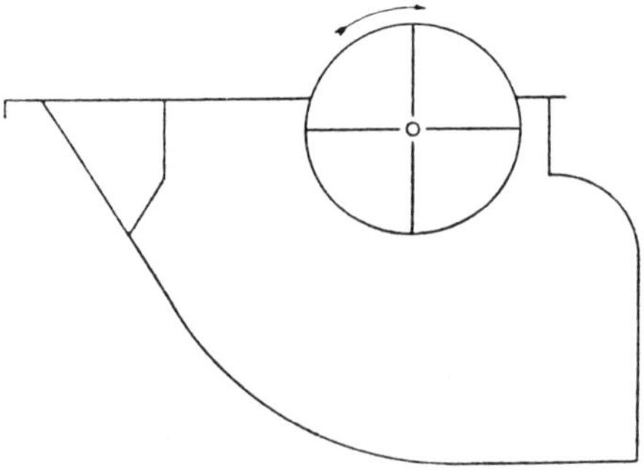

Figure 4.4 – Cross-section of overhead-paddle machine

lower degree of lateral circulation which may lead to uneven distribution of chemicals or dyes. Dyeing machines of the paddle type can be used for scouring and milling. Although such machines can be automated, this is seldom encountered in practice.

The Pegg Toroid and its later development the Pegg K-type machine (see Figure 4.5) are frequently used for scouring, milling and dyeing. The garments

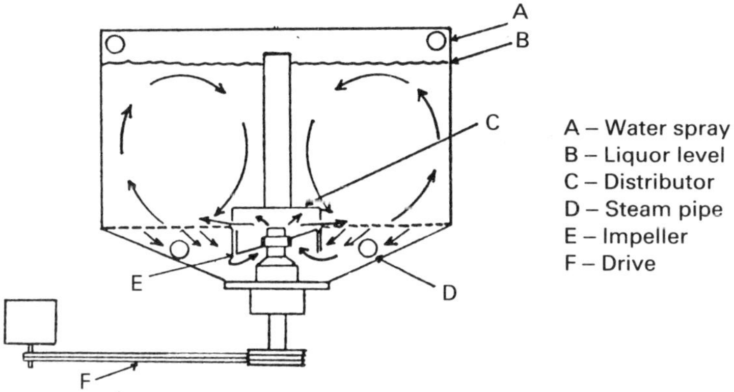

Figure 4.5 – Pegg K-type scouring and milling machine

circulate in the liquor in a toroidal path with the aid of an impeller situated below the perforated false bottom of the vessel.

Recent developments in dyeing machines have been concerned with processing garments at liquor ratios lower than the 30:1 usually associated

Figure 4.6 – Dytex machines

with the paddle machines. A number of machines have been produced including:

(a) Dytex machine (see Figure 4.6) – Neil and Spencer (UK)
(b) Milnor dyeing–extracting machine (marketed in different countries by a number of agents)
(c) Rotex machine – Pegg Whiteley Ltd (UK)
(d) Electrolux–Wascator Colorflex (Sweden).

These machines have a variable tumbler speed and can be used for scouring, milling, hydroextraction, dyeing and the application of shrink-resist treatments. They have multi-addition tanks and are usually highly automated. Liquor ratios of 10:1–15:1 are stated to be possible.

4.4.3 Drying Machines

In less sophisticated machines, where hydroextraction facilities are not available, the goods are hydroextracted in a centrifuge which may be of the low- or high-speed type. Final drying is carried out in tumble dryers which are

Figure 4.7 – Permac (Böwe) solvent processing machine

similar in appearance to the combined processing machines discussed in the previous section.

Conveyor-type radio-frequency dryers are sometimes used for hosiery and garments, following rotary hydroextraction.

4.4.4 Machines for Solvent Processing

The machines used for batch solvent processing are essentially developments from industrial dry cleaning machines. They consist principally of a rotating cage which is connected by means of pumps and associated pipework to various tanks for chemical addition and storage of solvent. All machines have two speeds: tumble and hydroextraction; solvent recovery and reuse is an essential feature. Machines are manufactured by:

(a) Böwe (West Germany) (see Figure 4.7)
(b) Donini (Italy)
(c) Neil and Spencer (UK).

These machines can be highly automated, often on a punched-card principle.

REFERENCE

1. Cockett, Wool Sci. Rev., No. 56 (1980).

CHAPTER 5

Services and Dyeing Machines for Piece Goods

5.1 SERVICES

All dyehouses require adequate services such as steam, compressed air, water and effluent disposal [1]. In wool processing the need for a supply of soft water must be emphasised, since scouring or milling in hard water will result in the formation of insoluble calcium and magnesium soaps. Such compounds cause considerable problems in subsequent dyeing and finishing. Compressed air is required to operate control equipment including pneumatic pressure devices.

The services should be supplied to the machinery by suitably installed pipework. The layout of the plant should be such that good flow lines are obtained without bottlenecks, usually following the order of processing (preparation, dyeing, finishing, warehouse), either in a straight line or in a U-formation, depending on the space available. Some problems may occur, especially when fabrics have to be dried more than once; the drying department should be situated centrally to accommodate the flow of goods from various sections of the factory. Single-storey buildings are preferable.

5.1.1 Automation and Control

Although the quality of milling and dry finishing is assessed subjectively, processes such as scouring, carbonising, dyeing and drying can be controlled automatically. As in the garment processing sector, however, the adoption of automation in fabric dyeing and finishing has been slow.

High-speed scouring machines can be controlled from a sophisticated panel, and drying machines can be monitored. Dyeing machines are capable of being automated to a high degree, and in many cases their control is not so complicated as in package or high-temperature dyeing methods.

Automation of the dyehouse has been summarised [1] and in the piece dyehouse the main areas for consideration have been identified as:

(a) Weighing of substrate
(b) Weighing and dispensing of dyes and chemicals, a mezzanine colour dispensary can be installed to advantage
(c) Control of dyeing machine operation
(d) Testing of dyes
(e) Selection of dyes
(f) Instrumental quality control of dyeing.

The ultimate aim is to adopt blind dyeing techniques, although this may be more difficult to realise in fabric processing in general and wool dyeing in particular. However, characterisation of batches in terms of dyeing characteristics will greatly assist in this objective.

5.2 PIECE DYEING MACHINES

5.2.1 Winches

One of the earliest and still most widely used dyeing machines for piece goods is the winch, the principle of the machine being illustrated in Figure 5.1.

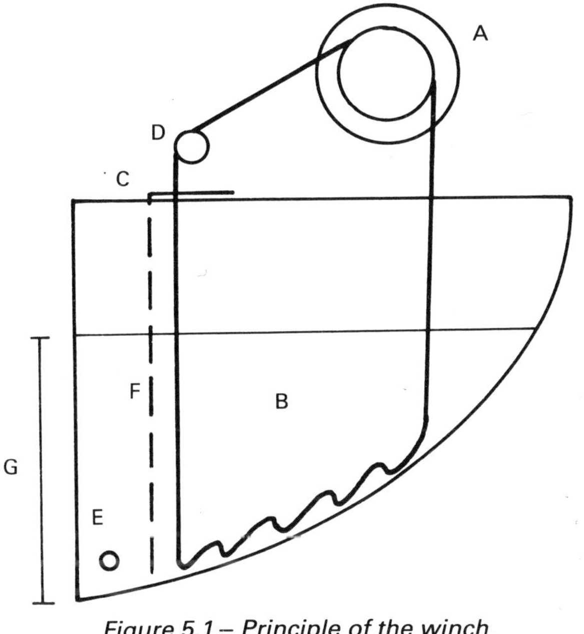

Figure 5.1 – Principle of the winch

Pieces are loaded over the winch reel (A), run into the liquor (B) and sewn end to end. During the dyeing process the ropes of fabric are run over the winch reel and collected in the relaxed state in the bottom of the vessel. They are then drawn through the peg rail (C) to prevent entanglement and over the jockey roller (D) back to the main reel. The dyebath is heated by the steam pipe (E) situated behind the perforated partition (F). Early machines were open but later versions were hooded to obtain better control at a higher temperature. Even so, atmospheric winches will not obtain dyebath temperatures much above 95°C. Early machines were constructed of wood but modern machines are manufactured from stainless steel.

The major developments in winch design can be summarised as follows:

(a) Use of shallow-draft machines (the distance G reduced) to give uniform temperature distribution throughout the liquor and reduce the drag on pieces from the dyebath to the jockey roller
(b) Elliptical or variable-geometry reels
(c) Driven jockey rollers
(d) Air jet at the jockey roller to assist ballooning
(e) Lowering of the main reel
(f) Lapping of the main reel to avoid fabric slippage
(g) Provision of two jockey rollers
(h) Variable machine speed, dependent on fabric type usually one revolution of the piece per minute or a fabric speed of 1 m/s
(i) Aids to liquor circulation, i.e. pump or impeller
(j) Device for loading one continuous length of fabric into the machine.

Most of the machine developments have been designed to improve the uniformity of treatment and to avoid the formation of running marks in the fabric, which is a serious problem with rope processing. Bagging of the fabric

Figure 5.2 – Conventional winch

(see Section 2.5.5) to promote ballooning helps to minimise this fault. The use of shallow-draft machines with the winch reel lowered so that the wet fabric does not approach it steeply from the jockey roller also reduces the incidence of the problem. Reduced liquor ratios can be employed for shallow-draft machines, but too low a liquor ratio will give rise to running marks.

High-temperature dyeing machines are seldom used for wool, with the possible exception of dyeing wool/polyester materials. A conventional winch is illustrated in Figure 5.2 and a shallow-draft version in Figure 5.3.

Figure 5.3 – Shallow-draft winch

In the traditional winch even level-dyeing acid dyes must be applied with care. Modern winches with improved control devices and dispensing facilities, as described above, have provided better regulation of temperature rise and hold periods, which are more critical for neutral-dyeing dyes. Nevertheless the increasing demands for higher fastness have encouraged wool piece dyers to evaluate other machine types.

5.2.2 Jet Dyeing Machines

From 1971 onwards there have been rapid developments in the field of jet dyeing machines for processing fabric at high temperatures in rope form.

These machines were originally designed to overcome the problems of dyeing textured polyester in conventional or high-temperature winches. Later the jet and overflow principles were extended to the dyeing of cotton knitgoods and blends at atmospheric pressure [2–5]. Atmospheric jet machines are less expensive than the original high-temperature jets.

Due to the relatively small market share of the total piece dyeing market that wool holds, there have been no specific developments in the jet dyeing machine field for wool. It was logical that jet dyeing machines should be considered for dyeing wool fabrics since replacements were needed for the old winches, and also a machine was required which would overcome the problems of winch dyeing, particularly the formation of running marks. The choice of a suitable machine proved to be difficult since by 1979 over 80 different jet machines were available on the market.

Holt and Harrigan [6] undertook a survey of jet dyeing machines and identified five basic machine types. These groups are summarised below.

Group 1

Fully and partially flooded jet dyeing machines with hydraulic fabric transport via a venturi nozzle. These machines, originally developed for textured polyester fabrics, generate a high velocity flow of liquor which has a harsh action on wool. Partially flooded machines have a pronounced tendency to generate foam. Design modifications to reduce the intensity of direct liquor action while maintaining a high liquor interchange in the nozzle have improved the suitability of these machines for dyeing staple-fibre fabrics including wool.

Group 2

Overflow machines with driven winch reel, a combination of hydraulic transport and driven winch. These are usually partially flooded machines and are considered to be of gentle action.

Group 3

Machines with a driven winch reel and a jet nozzle.

Group 4

Combined overflow/jet nozzle machines with a driven winch. These are normally partially flooded machines, regarded as offering considerable potential for wool piece dyeing.

Group 5

Machines using a form of mechanical conveyance to assist fabric transport in addition to a winch or jet/overflow system.

Following this survey, practical dyeing trials were carried out on eleven jet machines from groups 1, 3, 4 and 5. This work indicated that machines from group 4 offered the most potential for wool fabrics. There appeared to be a limitation on the weight of the woollen fabrics which could be processed satisfactorily on this type of machine, but all the worsted-type fabrics processed were satisfactory. Since fabric transport can be highly controlled by means of the overflow cascade, the adjustable nozzle pressure and the driven winch reel, it is possible to eliminate the formation of running marks and over-working of the fabric surface.

Investment in jet dyeing equipment is unlikely to be justifiable for a dyehouse specialising in wool, but these machines will be available in dyehouses processing fabrics made from synthetic-polymer fibres and their blends with wool. Figure 5.4 illustrates a jet machine which has shown much potential for dyeing woollen and worsted fabrics.

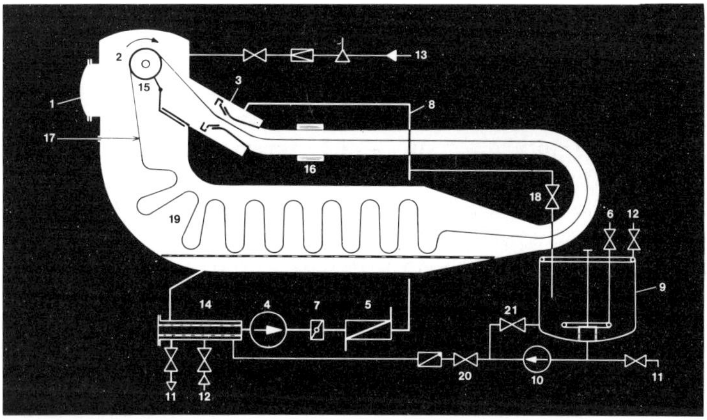

1 – Hinged cover
2 – Return pulley
3 – Injector with hydraulic seal
4 – Liquor pump
5 – Heat exchanger
6 – Steam connection
7 – Jet stream regulation
8 – Circulation system
9 – Preparation tank
10 – Addition pump
11 – Discharge valve
12 – Water connection
13 – Compressed-air connection
14 – Dyebath filter
15 – Cut-off valve
16 – Steam indicator
17 – Sample cutting device
18 – Back flow
19 – Fabric storage
20 – Connection valve to preparation tank
21 – Mixing valve

Figure 5.4 – Thenflow jet machine

Since the survey carried out by Holt and Harrigan little change has been made in jet dyeing machines. Probably the most significant modifications have been in the move to more even filling of the storage compartments of the machines, which has been achieved in a number of ways:

(a) Plaiting nozzles
(b) Split baskets to allow two lightweight fabrics to be processed in one tube
(c) Rotating baskets to avoid excessive tension on the fabric ropes as they are lifted from the storage chamber.

5.2.3 Beam Dyeing Machines

The beam dyeing process is a form of package dyeing in which the fabric is dyed in open width on a perforated cylinder. Atmospheric and high-temperature machines are available, the direction of liquor flow can be reversed and a high degree of automation can be achieved. The preparation of the beam of fabric must be carried out with care since any creases formed during winding or shrinkage on the beam become permanent during high-temperature treatment. However, when properly prepared, the fabric will not develop creases on the beam. Dyes must be carefully selected, since particles of precipitated water-soluble dyes or poorly stabilised metal-complex disperse dyes can be filtered out during dyeing.

Woven woollen and worsted fabrics are not normally beam dyed, for, apart from the relatively high cost of the machine and its ancillary requirements, there has been a fear that flattening of the fabric occurs. Beam dyeing has been more widely used for knitted fabrics, especially warp-knitted material. The beam dyeing method is extremely well suited for wool/polyester fabrics, when high-temperature techniques can be used. Air injection devices have been developed to reduce the tendency of fabric flattening. A beam dyeing machine is illustrated schematically in Figure 5.5.

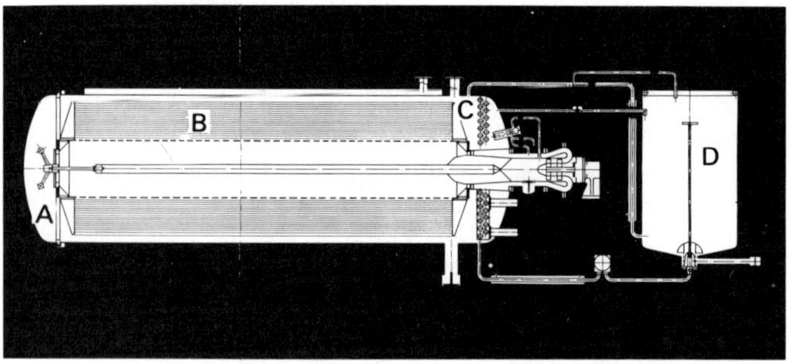

A – Main tank C – Steam coil
B – Beam D – Side tank

Figure 5.5 – Beam dyeing machine

5.2.4 Cold Pad–Batch Process

In general the runs of a given colour in wool dyeing are seldom sufficiently long to consider fully continuous dyeing as a viable technique. However, smaller batches can be handled economically by a pad–batch technique, carried out cold with reactive dyes [7].

The fabric is padded in open width by passing through a liquor containing dye, thickener, wetting agent and urea, excess liquor being squeezed out at the nip of the pad mangle. The impregnated goods are rolled up, or batched, covered with polythene sheeting to prevent drying out and stored for long enough to achieve maximum dye–fibre fixation. The batching time depends on the type of reactive dyes used and the depth of shade. After storage the material is washed-off in a dilute solution of ammonia, followed by thorough rinsing in water. The addition of sodium bisulphite to the padding liquor gives higher yields (with some dyes), improved levelness and a reduction in batching times.

Recent work has been concerned with reducing the storage time required for fixation by increasing the fabric temperature using radio-frequency (r.f.) heating. This would also allow the range of suitable dyes to be expanded to classes other than reactive types.

This method of dye application has not been widely used but a few units have been installed. The layout of a plant for the continuous cold pad–batch dyeing of wool fabrics is shown in Figure 5.6.

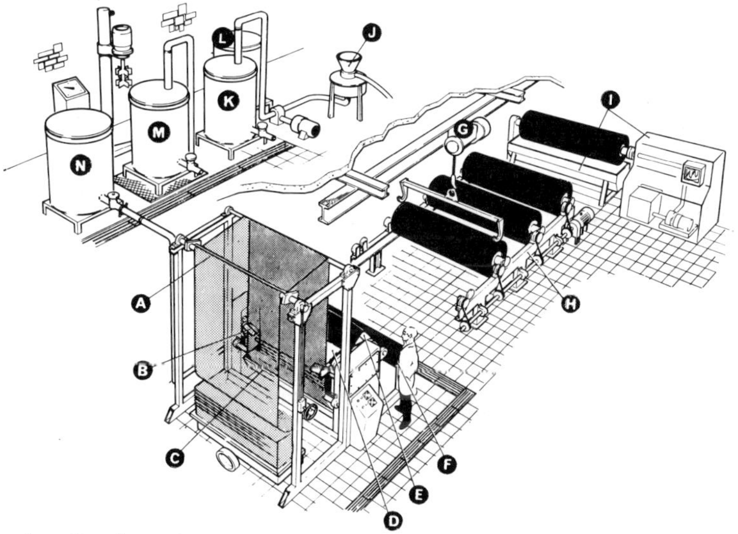

A – Tension rails
B – Centring guides
C – Pad trough
D – Mangle bowls
E – Curved opening roller
F – Batching roller
G – Transfer hoist
H – Batching creel
I – Beam washer
J – Mixer for thickener
K – Thickener storage
L – Urea stock solution
M – Dye make-up tank
N – Dye liquor storage

Figure 5.6 – Outline of cold pad–batch process

5.2.5 Choice of Piece Dyeing Machine

The winch is likely to remain popular for wool woven piece dyeing on cost grounds. Although atmospheric and high-temperature beams, jets and overflow machines are well established for dyeing other fibres and blends, they are unlikely to be available yet for wool fabrics, except as part of a multi-fibre operation. However, with the increasing amounts of washable woven fabrics coming onto the market that require piece dyeing to more stringent fastness specifications, dyes with better colour fastness have to be used. These dyes are more difficult to apply than the levelling acid dyes to obtain level dyeing, and the traditional winch machines are hardly suitable for this purpose. Dyeing machines with much improved liquor circulation and temperature control are desirable and this may provide an opportunity for the wider use of gentle-action atmospheric jet and overflow machines. So far, little use has been made of the cold pad–batch method.

Weft-knitted wool is seldom piece dyed; the winch would normally be used, but a jet or overflow machine, if available and suitable, would be preferred.

Since ITMA 1983 machinery manufacturers are reporting an upturn in demand for new machines, including piece dyeing equipment. Winches are likely to remain predominant, however, largely due to the higher costs of purchase of alternative machine types.

REFERENCES

1. Park, 'A Practical Introduction to Yarn Dyeing' (Bradford: SDC, 1981).
2. Ratcliffe, Rev. Prog. Coloration, **9** (1979) 58.
3. Marshall, Shirley Institute Publication S33 (1979).
4. Ratcliffe, J.S.D.C., **96** (1980) 94.
5. Simborowski, J.S.D.C., **96** (1980) 111.
6. Holt and Harrigan, IWS/CSIRO Publication (1979).
7. Wool Sci. Rev., Special Edition, Paris ITMA (1971).

CHAPTER 6

Drying Methods

After one or more wet treatments a fabric must be dried before storage or transfer elsewhere. This usually takes place in two stages by:

(a) A relatively cheap mechanical method to remove water from the interstices of the fabric

(b) Thermal treatment to evaporate water from within the yarns and from the interior of the fibres; this stage is much more costly in energy.

Handling problems limit the application of radio-frequency techniques.

6.1 WOVEN FABRICS

6.1.1 Mechanical Methods

The fabric is evenly distributed within a rotary hydroextractor or centrifuge, high- and low-speed versions being available. These will reduce the moisture content to around 25% (on dry weight) for wool fabrics.

An alternative method is to pass the fabric in open width through a mangle, although the moisture content is unlikely to be less than 45% after mangling. The method is less labour intensive, avoids the risk of rope creasing and allows the fabric to be plaited down mechanically for transfer to the next process. The advantages of hot mangling have been mentioned previously (Section 2.7).

6.1.2 Tentering

Thermal drying of wool consumes much energy and care must be taken to ensure that optimum use is made of that energy. With this in mind, the moisture content should be no more than 60% before tentering, and preferably much less. The finisher should attain this by centrifuging, mangling, passing over a suction slot or a combination of mechanical methods. The tenterman should measure the width of the fabric before tentering to ensure that no shrinkage problems will be encountered. It is a basic fact, often sadly forgotten, that the pre-tentered fabric is in the fully relaxed state. If the wet width at this stage is considerably less than the required finished width, then the finisher will have a problem with shrinkage on that particular fabric unless some form of permanent setting (i.e. crabbing or autoclaving) is carried out. It is normal to tenter woollen fabrics 5–9 cm wider and worsteds 5 cm wider than the required finished width, this extension being helpful with the removal of wet creases and to allow for some shrinkage during the dry finishing operations which follow. Approximately 10% overfeed should also be put into the fabric, but this will vary from one quality to another.

Modern tentering machinery has several features which ensure fault-free fabric drying and maximum utilisation of energy. These may consist of the facilities described below.

Weft-straightening Devices

These devices are linked to bow and skew rollers which operate automatically so that the weft threads are straightened. Automation is essential as manual control of weft straightening would be impossible at high speed. Slower tentering to permit manual control is likely to cause overdrying.

Control of Moisture Content

A device is available to enable the finisher to dry the fabric to a required moisture content, say 8%, and can be automatically wired into the machine drive so that the speed control is governed through the monitoring device at the fabric exit.

Control of Relative Humidity

The control unit is situated in the exhaust outlet and can either regulate the speed of the exhaust fan or govern the damper. This ensures that the exhaust being removed from the drying chamber is fully saturated.

Heat Exchanger

This utilises the heat from the exhaust to pre-heat the air being taken into the drying chamber.

It is important to make certain that too much moisture is not being removed from the fabric. The bound moisture from the fibre interior (below 8% moisture content) demands the maximum energy for removal. Many finishers are guilty of overdrying in tentering and then steaming and dewing to put moisture back into the fabric. A practical test is to observe whether the hairs on one's forearm stand on end when held close to the fabric leaving the drying chamber. If so the machine speed should be increased to avoid overdrying. The static electricity built up when overdrying is harmful; not only are excessive flock and dust attracted to the fabric, but interference with cuttling (folding) and machine performance can cause fabric damage.

A multi-layer tenter for wool fabric is shown in Figure 6.1.

Heat Setting

When heat setting wool/polyester blends it is advisable to install a moisture control device in the system before the fabric passes into the heat-setting zone. This ensures that it is dry before entry into the zone, thus enabling the

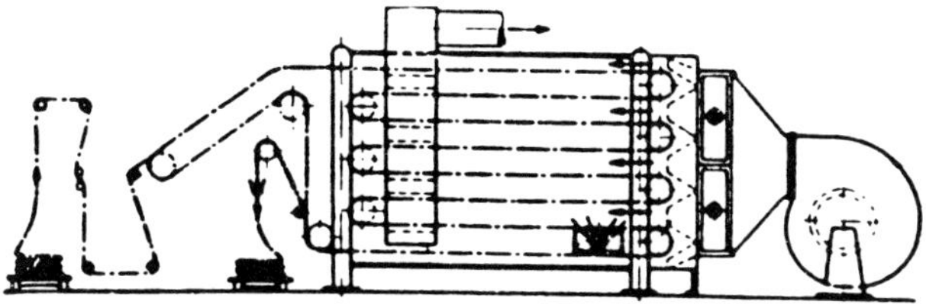

Figure 6.1 – Multi-layer tenter for wool fabric

heat setting of the polyester to take place at 170°C for about 30 s. Improved handle is achieved if the heat-setting width is slightly less than the drying width.

6.2 KNITTED FABRICS

6.2.1 Mechanical Methods

Interstitial water is removed by rotary hydroextraction as previously described.

6.2.2 Thermal Methods

Circular-knitted fabrics are thermally dried by passing them over a perforated tube through which hot air permeates, as shown in Figure 6.2.

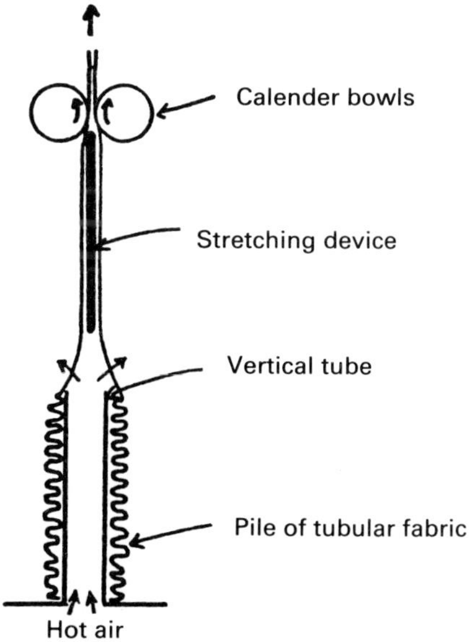

Figure 6.2 – Tubular drying machine

Alternative methods for circular fabrics are:

(a) Suction-drum dryer
(b) Brattice dryer
(c) Slit the fabric tube and stenter dry.

Flat-knitted fabric is dried on a stenter, similar in construction to a single-pass horizontal tenter. Spiral uncurling devices ensure that the selvedges are opened out and flat before the fabric is pinned.

CHAPTER 7

Mechanical Finishing

Wet finishing includes processes such as carbonising, scouring, milling and crabbing as discussed in Chapter 2. After dyeing and drying the fabric normally receives several dry finishing treatments which are described below. Various sequences of wet and dry finishing are considered in Chapter 8.

7.1 DAMPING

After drying the first process may be damping the fabric by either a water spray or steam to make it more receptive to mechanical finishing processes. Intermediate drying may be carried out between wet processes in an attempt to minimise creases or skewing of the weft. A typical damping machine is shown in Figure 7.1.

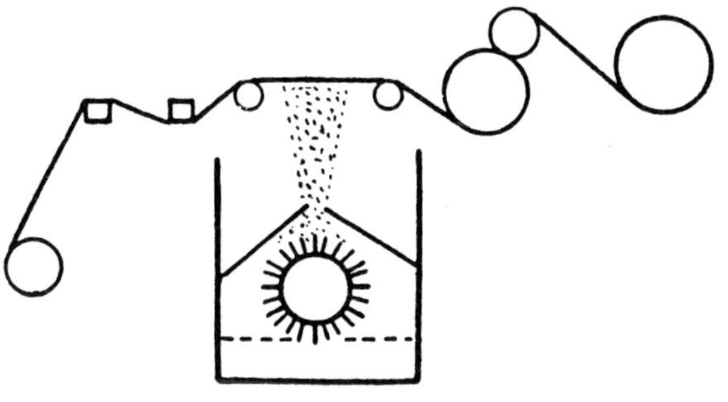

Figure 7.1 – Method of damping

7.2 RAISING

The appearance of wool fabrics can be altered dramatically by either raising or laying a surface pile. This takes place by passing the fabric over the surface of a cylinder which is covered with teasles or card wire. Great care is necessary during this operation as the raising action impairs the fabric strength.

Since the card-wire raising machine (see Figure 7.2) is more productive, it is customary to start raising on this machine and to complete the process on the teasle gig (see Figure 7.3). Certain effects can be obtained only by teasle action. The speed and effectiveness of raising depends on a large number of factors including machine settings, fabric tension, fibre length in the blend, yarn count and twist, fabric construction, previous processes such as milling, the presence of a softening agent, and the pH and moisture content of the

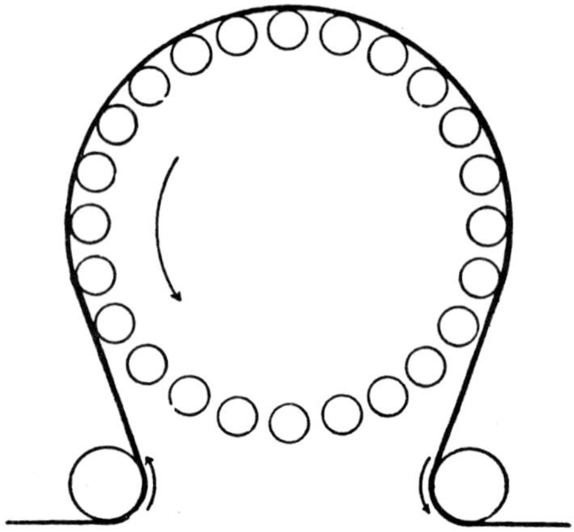

Figure 7.2 – Card-wire raising machine

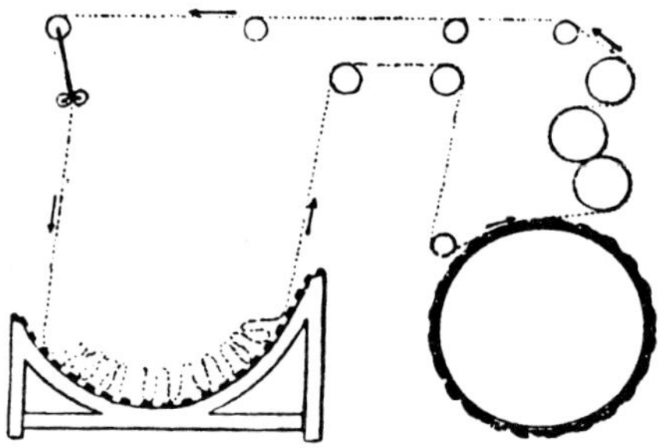

Figure 7.3 – Teasle gig

fabric. Raising should begin gradually to obtain a uniform finish. Moisture assists raising and the process can be carried out either wet or dry. In wet raising the effect is to lay the raised pile in the teasle direction on the fabric. Dry raising produces a more spongy effect on fabrics that have not been milled.

It is customary to raise after milling, however, so that raising takes place on a consolidated fabric. The warpway stress causes an increase in fabric length and shrinkage in the weft direction. Complete removal of some fibres from the material results in a weight loss, which must be taken into account when calculating the amount of milling required, and it is usual to give additional milling to allow for this weight loss when the fabric is to have a raised finish.

The weight loss incurred after wet raising is less than that after dry raising. For best results finishers will usually raise the fabric in the wet state as this gives most even raising, and has a gentler action than when performed dry. When producing fabrics that pass only one way over the raising barrel (i.e. layed or long pile) each piece should be examined carefully selvedge to selvedge and end to end, to ensure that the raising treatment is uniform.

7.2.1 Teasle Raising

This is generally accepted as the best way to raise woollen fabric. All other methods involve the use of metallic wire, which causes more damage to wool than the vegetable teasle. This difference in performance indicates that the fibres are plucked more gently from the weft by the teasle. The metallic devices, even with careful control, fail to imitate this gentle action.

Skill is required to set the teasles into the raising rods, and careful checking is necessary to ensure that no teasles remain proud, as this gives uneven and streaky raising. For best results the raising should be brought out of the 'bottom' of the fabric, by starting the operation with the older, used, teasles and then gradually replacing these with new teasle rods as the pile develops. Excellent, dense, laid-pile doeskins and billiard fabrics are produced in this manner.

7.2.2 Card-wire Raising

This is the artificial imitation of raising with the vegetable teasle. Initial capital cost is higher, but the durability of the metallic teasles lowers the running cost and improves the productivity and raising capability. The rate of pile production depends on the degree of drum contact. This machine is invariably fitted with a card-wire back roll to produce laid finishes. The metallic teasles are positioned on the drum surface at angles to each other thus reducing the possibility of streaky or uneven raising, and this produces a very slight 'diamond' effect which is a feature of metallic teasle raising. An excellent soft mouflon or velour finish can be obtained by pre-wetting the fabric and passing it twice over the drum with the opposite ends leading in turn.

Single- and double-action raising techniques are possible (see Figure 7.4).

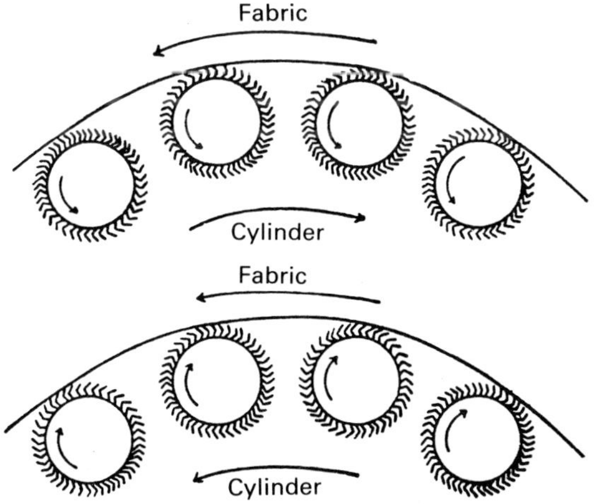

Figure 7.4 – Principles of single- and double-action raising

The single-action machine consists of 18 or 24 rollers covered with card wire and revolving in an anti-clockwise direction, whilst the drum in which they revolve travels clockwise. The fabric passes anti-clockwise over the surface of the drum or barrel so that one side of the weft yarn is raised. Several passes of the fabric over the barrel with the same end leading results in the formation of a laid pile, but it has a much fuller handle than that produced by wet teasle raising. By alternating the leading ends velour, mossy or velvety finishes can be achieved. Usually running speeds are about 15 m/min.

In double-action raising, as the name suggests, the machine can have twice the raising capability of the single-action machine. The raising barrel and the fabric run in an anti-clockwise direction whilst the pile and counterpile card-wire rollers run clockwise. The intensity of raising power of the pile and counterpile raising can be altered, and many different finishes can be achieved in this highly productive and very versatile machine, which can also incorporate a wire laying brush. Great care is required as the rate of raising on this machine can be so high that in unskilled hands a fabric can very quickly be over-raised. Fabric speed can be varied, but about 20 m/min is usual.

Although raising machines are normally used to produce raised or laid-pile effects, as a final finish, they can also assist in wet finishing processes. Blind finishes can be quickly obtained by scouring followed by wet raising, especially on lower-quality woollens and wool blends. The cover obtained by raising is milled into the fabric to give a dense blind finish. This may be the only method for such finishes on these goods, or on fabrics produced from highly twisted yarns. Even for higher-quality fabrics milling times can be greatly reduced by giving the fabric a light raising prior to milling.

With wool blends it is important to check the pilling performance of the finished fabric. If poor, the degree of raising should be modified to obtain satisfactory performance.

7.3 SUEDING

A slightly raised effect is obtained on wool fabrics by means of the sueding machine, which consists of a series of rollers covered with sandpaper or emery cloth. The degree of contact between the abrasive surface and the fabric can be varied using guide rollers.

7.4 BRUSHING

Fabrics that do not require dry raising are simultaneously steamed and brushed (see Figure 7.5) to lift the surface fibres before shearing. It is usually desirable to brush again after shearing to remove fibrous debris and restore the uniformity of the pile.

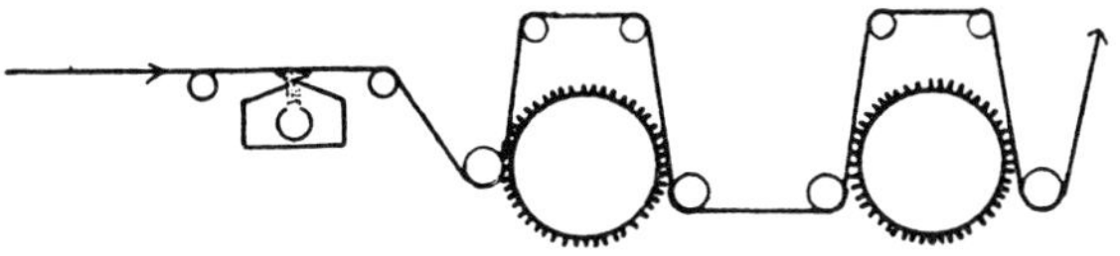

Figure 7.5 – Steaming and brushing machine

7.5 CUTTING/SHEARING

Shearing, cutting or cropping is carried out to trim and level the surface pile produced by raising and to remove surface fibres in clear-finish fabrics. The machine consists of revolving knives and an adjustable bedplate, as shown in Figure 7.6.

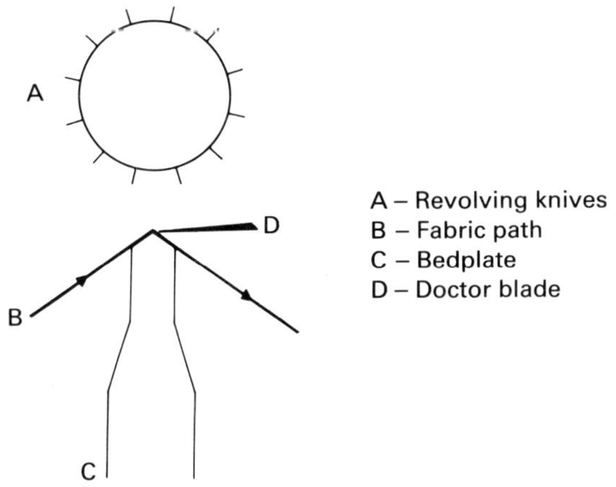

Figure 7.6 – Principle of a shearing machine

The back of the fabric is firstly sheared to remove knots and tails which have been pushed to the back during greasy mending and burling. If these were not removed the fabric passing unevenly over the bedplate would be damaged by the cutting blade. Shearing of the face of the fabric then follows, either once or twice. The height of the pile remaining is gauged by the finisher by feeling the tightness of the blade against the fabric. A close crop will result in a crisp stubbly handle and a high crop (hair end) produces a softer handle.

Normally a solid bed is preferred but hollow beds are available that will prevent cutting holes, should any knots be left on the back of the fabric. Modern solid beds have recessed edges so that very close cropping can take place without the risk of chopping the selvedge. It is important that the machine operative repeatedly compares both selvedge areas with the middle of the fabric for uniformity of pile, especially with colour-woven contrasts (e.g. white warp, dyed weft). It is prudent to pass the fabric through the machine a second time to ensure that both selvedge areas receive a similar cutting treatment.

Modern cutting machines are becoming very sophisticated, the following options being available:
(a) Metal detector to prevent damage to the blade or cylinder
(b) Automatic height of cut adjustment, whereby the finisher can select the height of pile required, which can be maintained irrespective of fabric thickness
(c) Sharpening of the blades without taking them from the machine
(d) Automatic location of the seam and movement of the bed or headstock to allow the seam to pass through
(e) Automatic oiling of the cylinder felts
(f) Variation of tension between cutting heads by the use of direct drives.

A cutting machine operates at 5–20 m/min and normally has one backing blade, a turnover and two face cuts, although other combinations are available.

7.6 WET SETTING

Laid-pile finishes should be wet set after the raised pile has been cropped to the required standard. The pile should be relaid to its original pre-cut state before setting. This process reduces the sensitivity of the pile to gentle friction and thus improves the durability of the finish.

The best results are obtained by potting in water at 75–80°C (see Section 2.6.2). The longer the immersion, the more durable the setting action. It is advisable to repeat the batching and immersion with the opposite end of the piece inside the roll to ensure uniform treatment. This is a slow and labour intensive operation, and therefore expensive; quicker methods can be used to set the fabric. The Hemmer Konticrab (Figure 2.16) is designed for continuous operation and gives durable setting of the pile. The wet fabric can be rolled straight from the raising machine and left to stand in this set state for some hours, but the effect is less durable than that of potting. The pre-cutting operation is most important for doeskins and billiards. If the residual pile is too long then an orange-peel effect will result which spoils the desired solid appearance. If the pile is cut too short then the resultant pile is stubbly and erect, and the lustre is very much diminished. Once more, great skill and experience is required from machine operatives.

7.7 NAPPING

After raising or cutting the fabric is drawn forward at about 4 m/min onto a stationary plush-covered table, above which is suspended a mobile pad covered with rubber or glass paper. Naps or beads of fibre are produced on the fabric surface by the action of the pad, this being controlled by eccentric cams to give a variety of effects. The fabric is treated discontinuously, in lengths of 0.5–1 m at a time. The weave in high-quality Eleysian and Crombie overcoats can be emphasised in this way.

7.8 PRESSING

Pressing is carried out on all clear-finish woollens and worsteds to make them firm, solid and smooth. This finish is often removed at the first stage of making up. In the hydraulic or vertical press the entire piece or a number of pieces are pressed simultaneously after cuttling mechanically, with electrically heated press papers inserted manually between the layers of fabric. A pressure of about 3–3.5 MPa (430–500 lbf/in^2) is applied and heating times vary from 10 min to 3 h. The fabric is then refolded so that the edges are in the middle and the process is repeated. A typical arrangement of presses and lifts is shown in Figure 7.7.

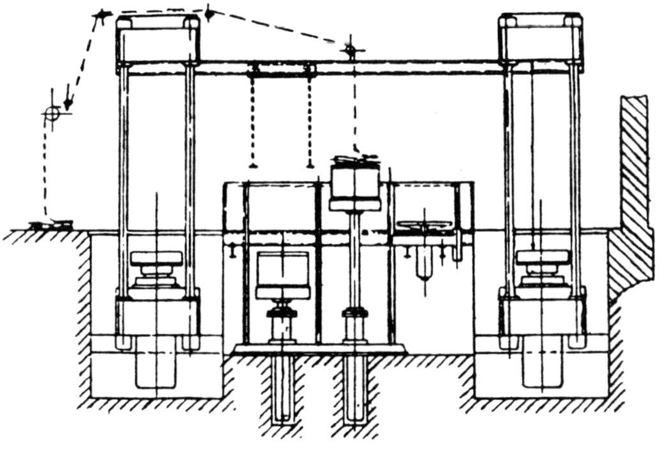

Figure 7.7 – Vertical press

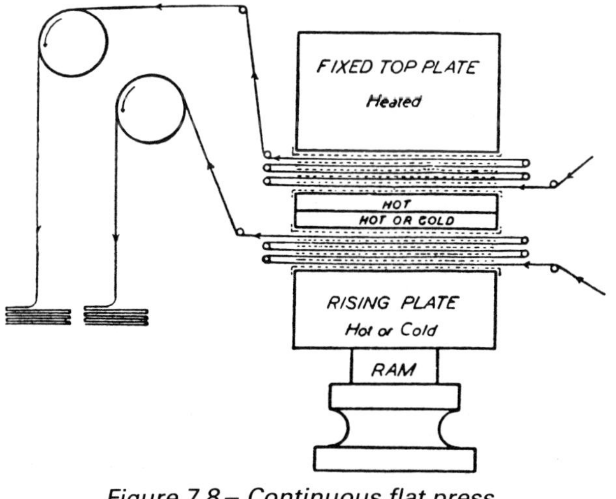

Figure 7.8 – Continuous flat press

An automatic flat press consists of a fixed top plate, an intermediate plate and a rising plate, all of which are heated. A ram below the rising plate provides the pressure (see Figure 7.8). The fabric is passed about five times to and fro between the plates via guide rollers, so that five thicknesses are pressed at one time. Pressure is applied when the rollers and fabric are at rest; after a given time the pressure is released and the plates separate. The cycle may be repeated four times per minute. The temperature and pressure should be adjusted to suit the fabric and the finish required. Creasing of fabric folds is avoided and, unlike rotary pressing, there is no stretching of the fabric.

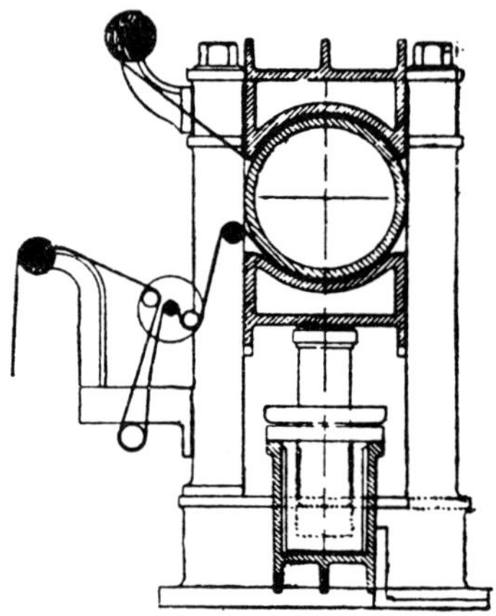

Figure 7.9 – Rotary press

The rotary press (see Figure 7.9) consists of a steam-heated cylinder which fits into the hollow of a steam-heated bedplate. The diameter of the cylinder is 40–60 cm. The fabric is carried through the press by rotation of the cylinder, and a pressure of 3.5–7.0 MPa (500–1 000 lbf/in^2) is exerted by pressing the bed against the cylinder. This machine operates continuously at about 5–10 m/min. Two-bed/two-cylinder versions are also available to give the fabric two pressings. The press may be provided with a belt or apron to allow for light pressing, thus giving fabrics some degree of stretch. Gentle steaming will remove any excess lustre or the papery handle associated with pressing. The various techniques for producing different effects are:

(a) Face to bed – high lustre
(b) Face to roller – lower lustre
(c) High pressure – more papery and firmer.

7.9 DRY SETTING

The main processes of this type are:

(a) Blowing with steam or decatising
(b) Autoclave or kier decatising.

Chemical means of obtaining set will be discussed in Chapter 10. The production of permanent press in wool fabrics and garments has been reviewed [1].

7.9.1 Steam Blowing

The final dry finishing process is blowing or pressing and is carried out to improve appearance, handle and drape, whilst at the same time removing the creases and crimps. The most common form of blowing consists of running the fabric onto a perforated cylinder between a cotton wrapper. Steam is then passed through the wrapper and fabric for 2 or 3 min and then the steam is removed by pumping. Appearance and handle depends upon controlled variations of wrapping tension, steaming conditions and pumping cycle.

To obtain satisfactory steaming or pressing it is important to ensure that there is sufficient moisture in the fabric being processed. It is advisable to prepare the fabric for blowing by dewing, covering with polythene and then leaving in this condition for 12 h so that the moisture is absorbed by the fabric.

The finish produced by steaming at atmospheric pressure is not permanent and some of it will be removed during making up and Hoffman pressing.

7.9.2 Autoclave or Kier Decatising

After making up the garment is pressed by passing steam through it at atmospheric pressure and this often eliminates much of the finished effect if the fabric has not been autoclaved after finishing.

As for steam blowing, the fabric is wound into a cotton wrapper to form a package, which is treated in an autoclave as follows:

(a) The autoclave is evacuated to a pressure of about 60 kPa (0.6 atm) and then steam bursts are introduced to pre-heat the package and create a passage for the steam through the fabric
(b) At the end of the evacuation cycle the steam is injected through the package until a predetermined pressure (70 kPa (0.7 atm)) is reached
(c) The steam is extracted from the package until the autoclave is once again evacuated to 60 kPa (0.6 atm), when the cycle is complete.

Woollen piece goods are generally steam decatised at atmospheric pressure to consolidate the fabric. This avoids collapse of the package during subsequent autoclaving, resulting in weftway pressure marks which are difficult to remove.

Appearance and handle can be varied by modifying the wrapper pressure, times of evacuation and steaming, and the direction of steam flow inwards or outwards through the package. Prolonged steaming with inwards flow at high wrapper pressure gives a durable lustrous finish. Conditions at the opposite extreme confer softer and less lustrous effects, but at low wrapper pressure the risk of package collapse increases.

The handle will greatly improve if the newly autoclaved fabric is left to condition in a cold damp atmosphere for 24 h.

7.10 FABRIC SHRINKAGE (RELAXATION) MACHINES

Relaxation shrinkage can be a problem during processing, particularly in making up, if fabrics have been overstretched in tentering or pressing. Traditionally the London shrinking process was applied at the making up stage to eliminate this relaxation potential, and several machines were developed for this purpose [2]. The fabric was suspended without tension between wet cloths and subjected to steam or a mist of water droplets in air.

7.11 KNITTED FABRICS

The processes discussed in the foregoing sections of this chapter have been described with woollen and worsted woven fabrics in mind. Many of these processes are applicable to knitgoods also, but these seldom require mechanical finishing. Milling to give a surface cover is important for knitted fabrics. Shrink-resist processes, which are discussed in Chapter 9, are exceedingly important for knitwear.

REFERENCES

1. Wool Sci. Rev., No. 46 (1973) 44; No. 47 (1973) 14.
2. Wool Sci. Rev., No. 8 (1952) 61.

CHAPTER 8

Finishing Routes

Individual steps in preparation, wet processing and dry finishing have been discussed in earlier chapters. Process sequences will now be considered for several typical fabric qualities (see Table 8.1). These, however, can serve only as guidelines since each manufacturer has virtually a unique method of obtaining a given fabric quality. The number and sequence of processes together with the duration or the number of repeat treatments of a given process will vary according to the manufacturer. Such variations have been discussed already for the preparation sequence (see Chapter 2).

TABLE 8.1

Typical Processing Routes

Woollen velour	*Woollen twill*	*Worsted/woollen*
Scour	Scour	Crab
Carbonise/bake	Carbonise/bake	Scour
Dry mill	Dry mill	Dye
Neutralise	Semi-neutralise	Tenter
Soap mill	Acid mill	Crop
Dye	Dye	Blow
Tenter	Tenter	
Knot	Knot	
Decatise	Steam	
Raise	Cut	
Cut	Decatise	
Felt		
Cut		
Steam		

Worsted serge	*Worsted gabardine*	*Cavalry twill*
Scour	Crab	Crab
Mill	Scour	Open-width scour
Dye	Dye	Dye
Tenter	Tenter	Tenter
Damp	Crop (twice)	Brush and damp
Crop (twice face, once back)	Steam	Crop (thrice face, once back)
Rotary press	Kier decatise	Blow
Kier decatise		

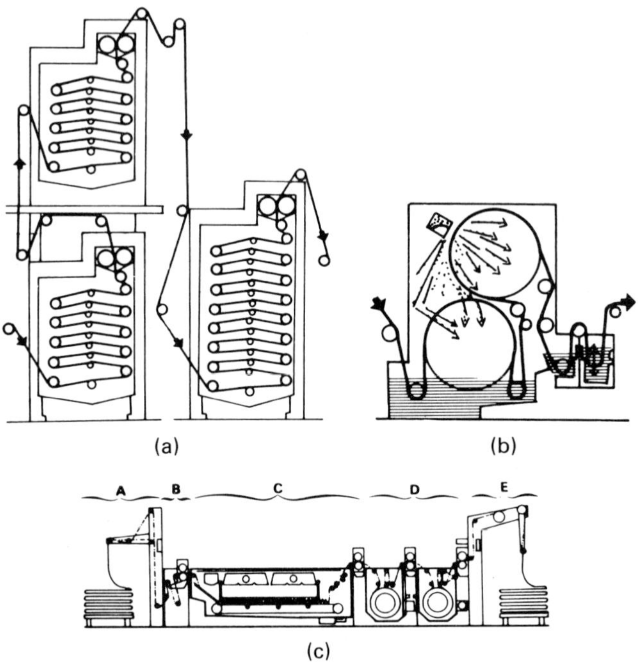

A – Entry feed
B – Pad impregnator
C – Dwell zone
D – Perforated drum with pulsed liquor flow
E – Final squeeze and plaiting

(a) Schematic view of Smith MC horizontal washing machine
(b) Representation of Küsters Roto-jet Cascade continuous washing and shrinking machine
(c) Layout of Hemmer Hydropuls system for continuous washing and shrinking of wool and wool-blend fabrics

Figure 8.1 – Continuous scouring ranges

8.1 PROCESSING ROUTES

It has often been stated that woollen fabric is made in finishing and worsted fabric is made in the loom. Thus the process sequences for woollens are usually longer than for worsteds, with knitgoods requiring the least finishing. The goods require a great deal of handling, not only in the numerous process steps (as in Table 8.1), but in the assembly of batches of various types for different machines:

(a) A number of pieces are sewn end to end to give a long continuous run, such as in carbonising

(b) Single pieces are sewn head to tail for rope processing, as in scouring, milling and dyeing
(c) As in (b) but for open-width processing, such as in scouring, raising and most dry finishing processes.

It is not surprising, therefore, that attempts have been made to develop continuous finishing methods for wool fabric processing. These can eliminate much of the handling, and open-width treatment avoids the running marks and creasing characteristic of rope processing. However, much longer runs are required, with greater control of fabric quality and standard width than in batchwise processing, where an average batch of five pieces is common. Continuous processing on modern equipment favours reproducibility and eliminates dye uptake variation arising in dye lots containing pieces from more than one carbonising or scouring batch.

8.2 CONTINUOUS FINISHING PROCESSES

Continuous processing of wool fabrics has been feasible for a considerable time, but the depressed economic climate of recent years has inhibited the necessary investment. Continuous open-width carbonising, scouring and crabbing (with drying if necessary) are followed by cold pad–batch dyeing and washing-off. Drying is by open-width suction extraction or tentering. The sequence of finishing processes, typically brushing, shearing, humidification and conditioning, pressing, decatising and shrinking, are all carried out in continuous open-width form. Carbonising, scouring (Figure 8.1), decatising (Figure 8.2) and dry finishing (Figure 8.3) of worsted fabrics in open width are well established, but traditional milling in rope form may represent the main obstacle to the adoption of a fully continuous sequence.

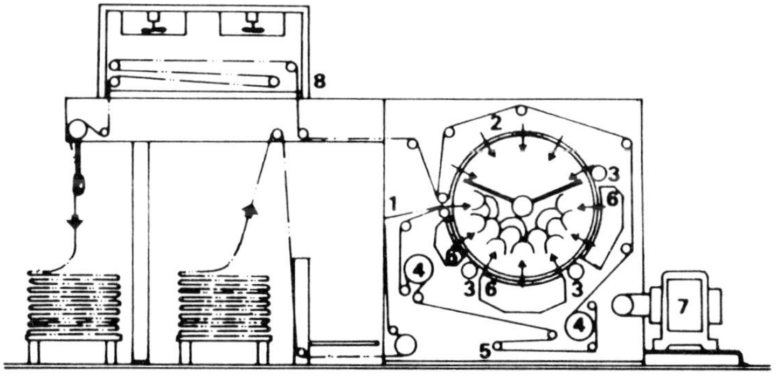

1 – Fabric tensioner
2 – Decatising cylinder
3 – Pressure rollers
4 – Heating rollers
5 – Felt-blanket tension roller
6 – Steam boxes
7 – Suction pump
8 – Cooling unit

Figure 8.2 – Hemmer Kontiset continuous decatiser

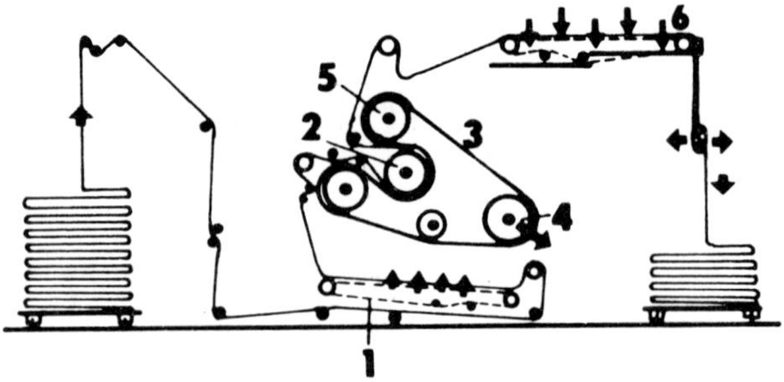

1 – Steam relaxing zone
2 – Heater cylinder
3 – Impermeable blanket
4 – Blanket-tension roller
5 – Belt drive
6 – Shock cooling zone

Figure 8.3 – Continuous dry finishing by the Hemmer Kontipress process

8.3 PROCESSING ECONOMICS

For continuous open-width operation a single-storey building is highly desirable. In many cases this would require drastic rebuilding of an existing batchwise dyehouse, or a new building similar in floor space to the old one. An economic study [1] indicated the following gains from this investment:
(a) 15% saving in capital equipment cost, although more extensive laboratory and testing facilities are required
(b) 50% reduction in labour
(c) 50% saving in steam and water
(d) 20% saving in electricity.

As already outlined, longer runs of standardised fabrics are necessary, but production rates exceeding 20 m/min should be possible. Creasing and similar problems should be eliminated.

Worsted-type fabrics may initially be best suited to this method of processing. However, with the economies and advantages that can be achieved, it is surely advisable to develop completely new fabrics, which, although similar to existing styles, can take advantage of the new continuous routes. A useful review of developments has been published [2].

REFERENCES

1. Wool Sci. Rev., No. 34 (1968) 10.
2. White, Textile Progress, **13** (2) (1983) 1.

CHAPTER 9

Shrink-resist Treatments

9.1 ORIGINS OF FABRIC SHRINKAGE DURING WASHING

When wool garments are washed in aqueous media dimensional changes can occur because of fabric relaxation and felting of the fibres. Relaxation effects occur when strains introduced into the material during yarn, fabric and garment manufacture are released when the garment is agitated in water. This release of strains occurs rapidly and is usually manifested as shrinkage, although in certain knitted structures extension may occur.

Whereas relaxation shrinkage can occur in fabrics manufactured from any fibre type, felting shrinkage is peculiar to wool. Felting shrinkage is caused by the relative movement of fibres within the yarn structure. When a wool garment is washed the mechanical forces imposed on the garment are transmitted to the individual fibres causing them to move slightly within the yarn. If this fibre movement were random, the effect on fabric structure would be negligible. However, the overlapping scales on the fibre surface exert a differential frictional effect, i.e. the friction between interlocking scale edges on adjacent fibres inhibits movement towards the tips and each fibre migrates preferentially in the direction of the root. Hence fibres are released from the interwoven yarn structure and become increasingly entangled with each other until no further movement is possible and a felt has been formed.

It is a simple matter to distinguish between relaxation and felting shrinkage. Relaxation changes are reversible, i.e. the garment can be temporarily pulled back to size and the surface appearance remains clear, although the weave or knit structure may be distorted. Felting is irreversible and is characterised by loss of surface definition and lower extensibility. Obviously the consumer wants a dimensionally stable garment, and it is necessary to control both relaxation and felting shrinkage in machine-washable products.

Relaxation shrinkage can be eliminated from fabric structures either by allowing stress relaxation to occur during wet and dry finishing, by processing under minimum tension or by stabilising the strains in the fabric using a high-temperature steam setting process. Techniques for controlling felting shrinkage are more complex.

9.2 CONTROL OF FELTING SHRINKAGE

Most commercial shrink-resist processes control felting shrinkage by either (a) modifying the scale edges to eliminate the directional frictional effect or (b) bonding the fibres together so that relative movement is not possible. Wool is an exceptionally versatile fibre and shrink-resist agents can be applied at various stages in the manufacturing process from fibre to garment. Shrink-resist treatments are most commonly applied to wool in the form of combed tops, knitwear (including socks) and woven fabrics (see Figure 9.1). Many processes have been described in the technical and patent literature [1], but only those of commercial importance will be considered here.

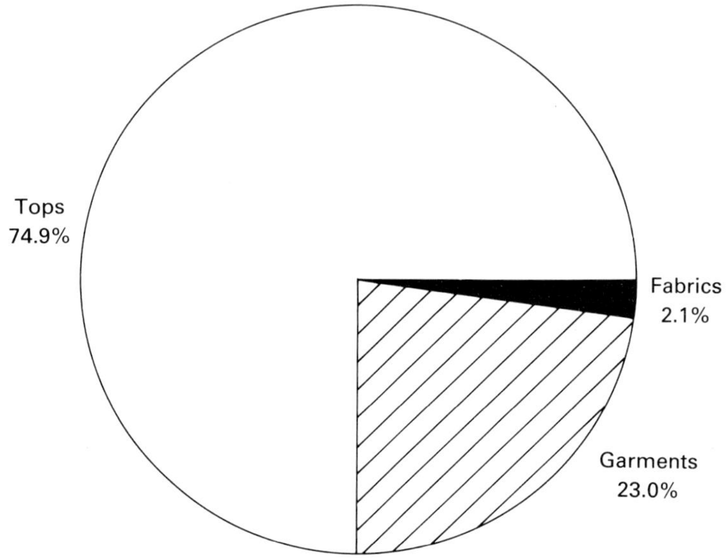

Figure 9.1 – Worldwide production of machine-washable wool in 1981 (23 500 tonnes wool treated)

9.3 CONTINUOUS TREATMENT OF WOOL TOPS

Wool tops are shrink-resist treated in a continuous manner using a modified suction-drum backwashing range with five treatment bowls [2, 3]. In each bowl the web of slivers passes around a perforated drum submerged in the treatment liquor. The liquor is circulated through the wool and the drum and then returned to the bowl by an impeller positioned inside the drum. This continual circulation ensures that the liquor penetrates the centre of each sliver so that all fibres are evenly treated.

9.3.1 Chlorine–Hercosett Process

The process most commonly used for the shrink-resist treating of wool tops is the chlorine–Hercosett process in which there are six distinct stages, as shown in Table 9.1.

The chlorination treatment, carried out at approximately pH 2, oxidises the surface scales and increases the anionic character of the wool. Finer-quality

TABLE 9.1

Typical Process Sequence for Continuous Chlorine–Hercosett Shrink-resist Treatment of Tops

Bowl No.	*Process*	*Treatment applied*
1	Chlorination	Available chlorine (1.5–2.0% o.w.f.) generated by mixing sulphuric acid and sodium hypochlorite
2	Antichlor/neutralisation	Sodium sulphite (1% o.w.f.) and soda ash to pH 8
3	Rinsing	Water
4	Polymer application	Hercosett 125 (16% o.w.f.)
5	Softener application	Cationic softener (0.2% o.w.f.)
6	Drying and curing	Wool dried to below 10% moisture content

wool fibres require a higher concentration of chlorine, because of their greater surface area and marked tendency to felt. The treated material passes into the next bowl, where the remaining hypochlorous acid is destroyed with alkaline sulphite solution. In the third bowl the residual chemicals are washed from the wool using an overflowing rinse.

The solution of the cationic polymer Hercosett 125 (Hercules) is metered into the fourth bowl. The agent is rapidly adsorbed by the oxidised fibre. A cationic softener, such as Alcamine CA New (Allied Colloids) is then applied to minimise fibre adhesion during drying, which would adversely affect subsequent drawing and spinning processes. Finally the wool is dried on a suction-drum drier and curing of the polymer takes place.

The chlorine–Hercosett process has a two-fold effect on the morphology of the wool fibres; the chlorination treatment reduces the depth of the scale edges so that the scale structure is less pronounced, and the polymer forms a layer that swells reversibly in water so that the scale edges become encapsulated. Hence the smooth fibres can no longer exhibit a directional frictional effect and felting does not occur.

9.3.2 Kroy Chlorination Technology

The most critical stage of the continuous chlorine–Hercosett process is the chlorination treatment because the reaction of chlorine with wool requires careful control. Inconsistent results will be obtained if the treatment is unlevel. The Kroy Deep-Im chlorination equipment [4] (see Figure 9.2) has been specially developed by Kroy Unshrinkable Wools Ltd (Canada) for chlorinating wool tops, and replaces the chlorination bowl in the normal suction-drum range.

The chlorinating agent is chlorine gas dissolved in chilled water, and this solution is sprayed onto the web of slivers just before it enters the U-shaped

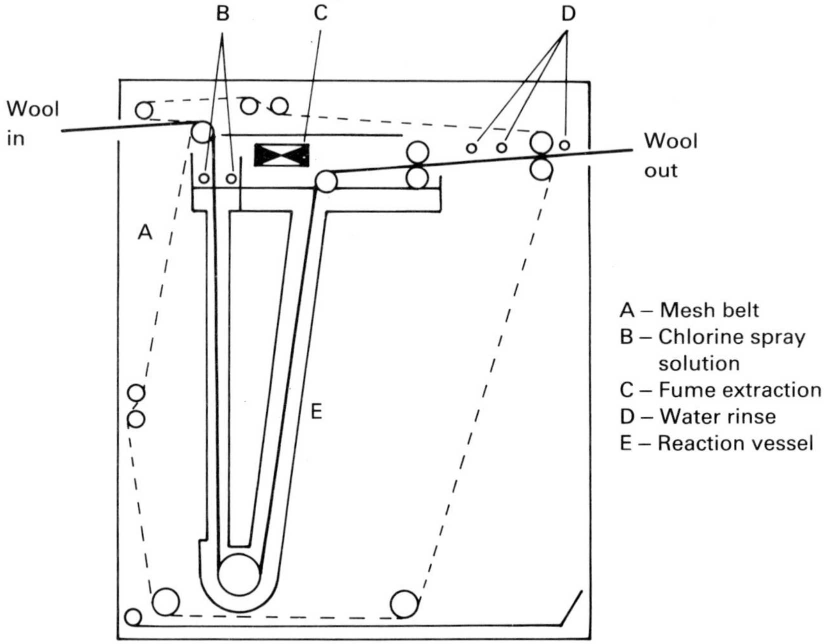

Figure 9.2 – Schematic diagram of Kroy Deep-Im chlorinator

reaction vessel. At the exit from the machine the slivers, which are carried through the machine between polypropylene mesh belts, are spray rinsed and then fed into the backwashing range for subsequent processing as described previously. The Kroy chlorinator gives a more uniform treatment than the suction-drum process, and lower levels of chlorination and polymer can be used. The success of the Kroy technology has inspired other organisations to develop improved chlorination techniques. At the time of writing, Fleissner (West Germany), Woolcombers (UK) and SAWTRI have reported the development of alternative chlorination processes with similar advantages claimed as those for the Kroy technology.

9.4 GARMENT TREATMENT

Wool knitwear is traditionally processed in aqueous liquors in side-paddle machines, although more recently rotary dyeing machines such as Dytex (Neil and Spencer), Dreher Milnor and Rotex (Pegg) have become popular because of the reductions in processing costs. A small proportion of knitwear is processed in solvent processing machines, which offer the advantages of automated dry-to-dry processing with low energy and water consumption.

9.4.1 Aqueous Processes

At present about 90% of all shrink-resist treatments for knitwear are based on chlorination. Originally acidified hypochlorite solutions were used but the rapid reaction of chlorine and hypochlorous acid with wool made the process difficult to control. Excessive treatment results in dye decomposition accompanied by extensive yellowing and loss of strength of the wool.

Techniques for controlling this reaction have been devised but, because of the instability and unpleasant nature of concentrated hypochlorite solutions, alternative sources of chlorine were sought. The chlorine generators most commonly used in aqueous shrink-resist processes are alkali-metal salts of dichloroisocyanuric acid (DCCA) (Figure 9.3), e.g. FiClor (Fisons) and Basolan DC (BASF).

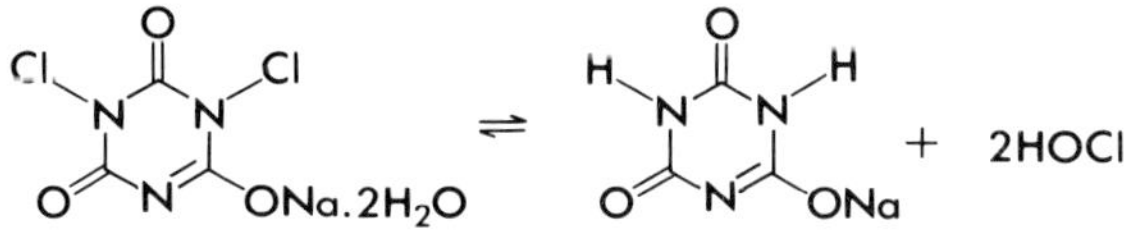

Figure 9.3 – Sodium salt of dichloroisocyanuric acid (DCCA)

In acidic solutions these products are capable of releasing approximately 35% of their own weight of available chlorine. However, in the presence of wool the salt may react directly with the wool rather than first go through a decomposition stage in which chlorine or hypochlorous acid is formed, which subsequently reacts with the fibre. Typically wool knitwear is chlorinated with DCCA at pH 3.5–4 and 20–25°C. These two parameters are used to control the reaction with wool; higher pH values and temperatures tend to accentuate yellowing of the fibre, and the reaction is excessively rapid at lower pH levels.

The general procedure for the shrink-resist processing of wool knitwear is shown in Figure 9.4. Modifications of this procedure incorporate potassium permanganate, which is claimed to give less yellowing of the wool and a softer handle.

After chlorination residual chlorine in the bath and chloramines on the wool must be destroyed by treating with a reducing agent such as sodium sulphite. At the same time the wool is neutralised, otherwise sulphurous odours become evident during storage. During this treatment most of the yellowing that occurred during chlorination is cleared. It is good practice to rinse after the antichlor treatment and it is essential with deep colours to use some form of alkaline backscour to remove dyed wool debris. This may become solubilised during chlorination and would otherwise lead to poor colour fastness if allowed to remain on the fibre.

Where a high level of machine washability is required then it is usually necessary to apply a polymer in addition to the chlorination process; polymers used include Polymer G (Precision Processes (Textiles) Ltd), Hercosett 125 and Basolan SW (BASF). All of these polymers are cationic and readily exhaust from neutral or mildly acid baths onto the anionic surface of chlorinated wool.

Scouring ↓	
Milling ↓	
Chlorination ↓	DCCA, 2–5% o.w.f. pH 3.5–4.0 40–60 min at 20–25°C
Antichlor/neutralisation ↓	Sodium sulphite, 5% o.w.f. Soda ash, 3% o.w.f. 20 min at 30°C
Rinsing ↓	(Heavy depths at pH 8–ammonia solution) 5 min at 50°C
Polymer application ↓	Hercosett 125, 16% o.w.f. or Basolan SW, 3% o.w.f. or Polymer G*, *x*% o.w.f. pH 4–6 20 min at 25–30°C
Softener application ↓	As required
Hydroextraction ↓	
Drying	

* Polymer G is used in the Dylan GRB process, details of which are confidential to Dylan

Figure 9.4 – Typical sequence for chlorination–shrink-resist treatment of garments

Basolan SW is a particularly effective shrink-resist agent which requires a lower degree of chlorination to achieve similar levels of performance.

If necessary, a low concentration of a cationic softener can be applied to the polymer-treated garments, although excessive concentrations may cause problems of pilling on some products. The mechanism of action of chlorination–polymer processes applied to garments is similar to that for continuous top treatment except that, in the case of the garments, the polymer acts as an adhesive between adjacent fibres and this further restricts fibre movement.

9.4.2 Non-chlorination Processes

Although chlorination–polymer processes are widely used, they suffer from the serious disadvantages that they cause yellowing of the wool and dye decomposition, resulting in colour changes and a deterioration in wet fastness of the dyeing. To overcome these problems non-chlorination processes have been developed for processing wool knitwear, based on the application of reactive pre-polymers to the wool which are subsequently polymerised on the fibre surface. The two products in commercial use are Nopcolan SHR3 (Diamond Shamrock) and Synthappret BAP (BAY). The systems are, at present, only used to a limited extent and consequently will not be described in detail here. Techniques and mechanisms of application have been discussed elsewhere [5, 6].

9.4.3 Solvent-based Processes

As previously mentioned, there are certain advantages to the processor in using solvent processing equipment for finishing knitwear, and consequently it has been necessary to devise shrink-resist processes that can be applied from this medium. Several processes, based on the use of solvent-soluble reactive pre-polymers, have been described in the literature [5, 7, 8], but the main difficulty is finding a polymer system that cures on the wool garments and not on the cage of the machine.

The most successful system to date is the DC109 process (Dow Corning) which is based on silicone chemistry. This is a two-pack system comprising a poly(dimethylsiloxane diol) and a low molecular weight alkoxy silane compound which, in the presence of atmospheric moisture, hydrolyses and reacts with the diol to form a cross-linked polymer. Since solvent processing can be carried out under essentially anhydrous conditions, polymerisation does not occur until the garments are removed from the machine and exposed to the atmosphere.

The main stages in the process are:

(a) Saturate garments in resin solution
(b) Extract to 100% liquor retention
(c) Spray catalyst solution onto garments
(d) Tumble garments for 10 min
(e) Dry garments.

The polymer solution is stored in a tank in the machine and its concentration is adjusted to give a treatment level of 3.5–4.0% o.w.f. (depending on nature of garments) at a liquor retention of 100% after centrifugal extraction. The solution of catalyst is then sprayed onto the garments as they tumble in the cage of the machine, at a catalyst application level of 0.4–0.44% o.w.f. (i.e. one part catalyst to nine parts polymer). The garments are tumbled for a further 10 min to allow the catalyst and polymer solutions to mix and then dried.

On exposure to moist air the silicone elastomer is formed on the fibre surface, with bridges of polymer between adjacent fibres. These bridges bond the fibres together so that no migration is possible during washing but, because the polymer is a soft silicone elastomer, the handle of the garments is not greatly impaired. Pilling is reduced because fibre movement is inhibited. The hydrophobic film prevents subsequent overdyeing, but the wet fastness of dyed garments is significantly increased.

9.5 FABRIC TREATMENT

In the past it has not been common practice to machine wash tailored wool garments and, accordingly, the development of washable woven fabrics has been slow compared to that of knitwear. With changes in the consumer's attitude to machine washing there is a growing interest in producing washable woven fabrics [9–11] and three processes are in commercial use. These are all based on the application by padding of reactive pre-polymers, which are subsequently polymerised on the fabric to form a network of fibre–fibre bonds within the yarns which prevent fibre migration occurring during washing. The treatments are applied to dried scoured fabrics of approximately neutral pH.

9.5.1 Synthappret BAP

This process involves application of the reactive pre-condensate Synthappret BAP together with the polyurethane dispersion Impranil DLH (BAY); the pad liquor formulation for treating a woven wool fabric is:
- Synthappret BAP, x g/l
- Impranil DLH, x g/l
- Sodium bicarbonate, 3–5 g/l.

The polymer concentration applied ($x = 10–30$ g/l) is determined by the fabric structure and liquor pick-up in padding. The sodium bicarbonate acts as a curing catalyst and during drying decomposes to the carbonate with a consequent increase in pH, which initiates the polymerisation of the Synthappret BAP. The liquor pick-up during padding is usually within the range 60–80% and the fabric must be heated to a temperature of 160°C to ensure curing is complete. On certain fabrics the polymers may produce a harsh handle, and if subsequent dry finishing does not produce an acceptable handle it may be necessary to incorporate a wash-off treatment in the process.

9.5.2 Ultratex ESB

Ultratex ESB (CGY) is a silicone elastomer system which imparts a very soft handle to treated fabrics. There are two components in the system, a silicone polymer (Ultratex ESB) and a cross-linking agent (Ultratex Catalyst EW) which are applied in the ratio 80:1. A formulation, applied at an expression of 80%, that would produce a machine-washable finish on most fabrics is:
- Ultratex ESB, 125 g/l
- Ultratex Catalyst EW, 1.5 g/l.

After padding the fabric is dried, batched and cured for several days at room temperature.

9.5.3 Nopcolan SHR3

Whereas the other products described are applied by pad–dry processes, Nopcolan SHR3 is applied by a pad–batch technique. The fabric is padded at 100% expression with a liquor containing:
– Nopcolan SHR3, 100 g/l
– Sodium sulphite, 20 g/l
– Thickening agent, 8 g/l

and stored for 16 h at room temperature in rotating batches, during which time polymerisation occurs. The fabric is then washed-off and dried. Important features of this method of treatment are the soft handle produced and good smooth drying properties of the fabrics after washing, possibly due to a flat-setting effect of the sodium sulphite.

REFERENCES

1. Rachel Makinson, 'Shrink-proofing of Wool' (New York: Marcel Dekker Inc., 1979).
2. Lewis, Wool Sci. Rev., No. 54 (1977) 2.
3. Lewis, Wool Sci. Rev., No. 55 (1978) 33.
4. Byrne et al., Textile Horizons, **4** (8) (1984) 24.
5. Cockett, Wool Sci. Rev., No. 56 (1980) 2.
6. Cockett et al., J.S.D.C., **96** (1980) 214.
7. Cockett and Kettlewell, Text. Month, (May 1979) 63; (June 1979) 62.
8. Greenwood, Knitting International, **84** (Jan 1977) 99; **84** (Feb 1977) 66.
9. Allanach et al., Proc 6th International Wool Text. Res. Conf., **5** (1980) 61.
10. De Boos and White, J. Textile Inst., **69** (1978) 194.
11. Lewis, Text. Research J., **52** (1982) 580.

CHAPTER 10

Chemical Finishing

Little chemical finishing, as it is understood in the cotton industry, is carried out on wool fabrics, which already possess many desirable intrinsic properties. The wet and dry finishing processes described in earlier chapters are adequate for most requirements. Shrink-resist treatment is the main chemical modification in wool finishing. In terms of weight of fibre processed, mothproofing is an important treatment, but about 90% of this is for the production of carpets, treatment usually being carried out at the loose-stock or yarn dyeing stages.

10.1 SETTING

As described earlier (see Sections 2.6 and 7.9), wool can be given a durable set by crabbing, potting and blowing. Setting is an essential prerequisite for easy-care and washable fabrics. Chemical setting can be achieved by treating the wool with a reducing agent able to break the cystine cross-links in keratin. These can be re-formed in a new structural arrangement after setting. Steaming, or alternatively the addition of an agent that can also disrupt the interchain hydrogen bonds in keratin, is essential. In practice chemical setting is now less important than formerly since it has been largely replaced by pressure decatising for flat-set and by Lintrak for crease-set textiles.

The use of sodium bisulphite was the basis of the Immacula process [1], and other processes have been developed [2] using agents such as ammonium thioglycollate and tetrahydroxyphosphonium chloride (THPC). These chemical setting processes were used in conjunction with a shrink-resist treatment. A typical processing sequence is:

1. Scour and mill
2. Shrink-resist treat
3. Dry
4. Set.

A typical flat-setting process is to pad the fabric to give an uptake of 50% by weight of a 1% solution of sodium bisulphite at pH 4.5–6, followed by immediate steaming for 5 min on a blowing machine at 0.5–0.6 MPa (5–6 atm) pressure [3].

Monoethanolamine sulphite (or sesqui-sulphite) has been used to set durable creases in trousers and skirts by spraying methods, followed by steaming; careful dye selection is required.

10.2 FLAME-RESIST TREATMENTS

Although wool is intrinsically flame retardant, further flame-resistant properties are necessary for end uses such as aircraft carpets, furnishings and protective clothing for firemen or racing drivers. The Zirpro processes, available under licence from IWS, are based on the use of compounds such as potassium fluorozirconate. Low smoke emission treatments are gaining importance for aircraft use, and with these chrome dyes should not be used. Careful dye selection is very important. Several 1:2 metal-complex and milling acid dyes give the fastness required for flame-resistant contract carpets and furnishings. The flame-resistant treatments are often best applied by exhaust or padding methods as the final stage in dyeing or wet finishing.

10.3 MOTHPROOFING

Wool is susceptible to attack by moths and the carpet beetle, so that insectproofing agents are often applied, particularly for carpets. The process can be carried out during dyeing. Several products are commercially available, but they are not all equivalent in protective effect. Typical applications are 0.6% Eulan U33 (BAY), 1% Eulan WA New (BAY) and 1–1.5% Mitin FF High Conc (CGY).

Apart from carpets, the only substantial need for mothproofing is on upholstery fabrics, blankets and uniform fabrics. The Eulan WA New treatment mentioned above is satisfactory for knitwear but is seldom used nowadays. Highly effective products based on dieldrin were popular in the 1960s but these are now environmentally unacceptable and, in Australia, dieldrin-resistant moth strains have developed.

Finishes based on the application of only 0.02% o.w.f. of a pyrethroid-type agent have been introduced recently [4, 5]. These products can be applied from the dyebath. Eulan WA New, Eulan U33 and most pyrethroid formulations can also be applied as an aftertreatment following scouring or milling. Treatment is carried out at pH 4–5 at 40–50°C for 15–20 min.

10.4 SHOWERPROOFING

Heavily milled wool fabrics are to some extent resistant to rain, but no fabric can be completely water resistant unless coated with rubber or a similar substance. This is not done for wool since it impairs the fabric's appearance and comfort properties, but various chemicals can be applied to increase the showerproof properties.

Paraffin wax dispersions have been applied to wool fabrics by padding followed by a solution of aluminium acetate or formate at 50°C. A protein-based product, such as Mystolene KP (Catomance), can also be used, this being diluted and mixed with aluminium triformate immediately before use. This product is substantive to wool and so can be applied by an exhaust method using 3% of the product at 50°C, treatment being continued until the bath is clear.

More sophisticated treatments are based on the use of silicone compounds such as polydimethylsiloxanes, and these are best applied by padding techniques followed by curing. Organotitanium compounds have also been employed. The fabrics are impregnated, hydroextracted and dried but then require 10–14 days cold storage for curing, during which time polymerisation is completed.

10.5 SOFTENING

Handle modifiers are seldom applied to wool since the fabric has its own special handle, often enhanced by wet or dry finishing processes. Softeners are applied to knitwear, especially to shrink-resist-treated materials as this process can harshen the handle. Such processes have been discussed in Section 4.1. Softeners are often applied to fabrics from the dyebath prior to raising to act as a raising assistant. An acid fabric pH assists raising. Products such as Ceranine HCS (S) are used at a level of 1–2%.

REFERENCES

1. BP 775 486.
2. Bajpai and Whewell, J.S.D.C., **77** (1961) 350.
3. Farnworth, Lipson and McPhee, J. Textile Inst., **51** (1960) T1504.
4. Shaw and Shepley, J. Textile Inst., **72** (1981) 92.
5. De Sousa et al., J.S.D.C., **98** (1982) 79.

CHAPTER 11

Inspection, Quality Control and Fabric Properties

11.1 FABRIC INSPECTION

Fabrics are examined for faults and as a general check on quality at various stages of the manufacturing sequence. In the case of pieces this is usually carried out by passing the fabric over an inspection table set at an angle of 45°. Fabric passage is assisted by rollers and a cuttling device is usually installed to refold the fabric after examination. The table normally has a glass top under which light tubes are mounted so that the fabric can be inspected by transmitted or reflected light.

This perching operation can be carried out at various stages of the process, but greasy inspection after fabric manufacture is almost always carried out. Weaving or knitting faults can be identified and if possible corrected by mending, thereby increasing the quantity of saleable fabric.

Perching is also carried out as a general rule on the finished fabric, and at this stage little corrective action can be taken without submitting the piece to reprocessing. Faults such as holes and stains are usually tagged or marked with strings and there is a commercial limit to the number of strings allowable in a sales piece, normally seven per 100 m.

Inspection may also be carried out at intermediate stages. For scoured pieces, which are raised before dyeing, perching is sometimes carried out after raising and before passing the pieces to the dyehouse. Perching may be carried out after dyeing as a check on shade and levelness. The weight and dimensions of the piece can be checked during intermediate perching to confirm fabric yield and for specification testing.

11.2 TESTING AND QUALITY CONTROL

Various testing procedures can be applied throughout the dyeing and finishing operation to check the quality of the goods at the various stages and to see that production is up to specification. Such testing is also necessary in the development of new products and processes. Many of these tests have been adequately described elsewhere [1] and many have reached the status of British Standards [2]. The object of this section is not, therefore, to discuss the test methods but to give an indication of the tests carried out and in some cases the acceptability of results.

11.2.1 Work in Progress

Weight per unit area of fabric is determined at a number of stages of the processing sequence; this assists in deciding whether or not the fabric conforms to specification. After carbonising the acid content and uniformity of distribution across the piece is important. The acid content of the fabric after water rinsing must be known to allow for correct calculation of alkali concentations in scouring.

The pH, oil content and total extractable impurities, are determined after scouring; satisfactory values which will give acceptable fabric for dyeing are:

(a) Oil content, not above 0.5%
(b) Total extractables, not above 1.7%
(c) pH of aqueous extract, 7.5–9.0.

The pH of the fabric as it is received for dyeing is an important parameter affecting level dyeing and must be adjusted according to the class of dye being used. Levelness and fastness properties must be checked after dyeing to the desired colour match. Blind dyeing techniques are adopted if feasible. Methods are available for the instrumental assessment of both colour and levelness (from end to end and across the width) using single-number shade passing techniques [3, 4]. Allowance must be made for any colour changes likely to arise in finishing (e.g. raising or cropping). Thus target colour standards are usually required to correspond with the process stages at which colour assessments are made.

The fabric pH prior to raising will affect the efficiency and reproducibility of the process. The moisture content of the fabric after drying should be monitored, preferably continuously by one of the systems that can be fitted to the tenter. Both over- and under-drying should be avoided.

Control can be exercised in bleaching and shrink-resist processes over the chemical composition of the baths. After finishing the shrinkage potential can be determined. Relaxation shrinkage is determined by steaming in the Wira steaming cylinder, when shrinkage should not be more than 1–2% in both warp and weft directions. Felting shrinkage is determined in the Cubex or Wascator machines for shrink-resist-treated materials; area felting shrinkage should not be more than 10% in a 3 h test.

The results of many dry finishing techniques such as raising and cutting can only be assessed subjectively by comparing the batch in process against previously accepted standard patterns.

11.2.2 Finished Products

Many purchasing organisations such as chain stores, users of materials for contract end uses or of high specification fabrics require that finished fabrics conform to well defined specifications. Such finished product testing is essential. The design of a suitable form for specifications has been given [1], but the properties tested would include those listed below, depending on fabric type and intended use:

(a) Moisture content
(b) Fibre content
(c) Weight per unit area
(d) Felting shrinkage
(e) Relaxation shrinkage
(f) Fabric strength
(g) Abrasion resistance
(h) Pilling
(i) Flammability
(j) Showerproofing
(k) Mothproofing
(l) Fastness to:
  Light
  Water
  Washing
  Perspiration
  Wet and dry rubbing.

11.3 FABRIC PROPERTIES

Changes in fibre quality, blend composition, or spinning or fabric-forming specifications will cause significant changes in the properties of the fabric. This will have serious results if the standard sample has been manufactured using differently specified materials from subsequent bulk. Similarly changes in processing sequences, times, etc. will cause changes in the finished product. There is thus a very strong argument for standardisation of materials and processes. This is not always possible since the results of some finishing treatments can only be judged subjectively, but nevertheless standardisation of many aspects is possible.

The severity and duration of processing obviously affect properties such as fabric strength. Alkaline treatments are generally harmful to wool and treatments such as dyeing are best carried out within the pH range 4.2–4.8. Processes that modify the fibre, such as shrink-resist treatments, must be carefully controlled.

Thus it is essential that standard methods and material specifications be established as far as is possible. Routine tests can be carried out to ascertain the properties of standard acceptable materials. From this the effect of changes in material or process on each parameter can be determined. For example, increasing the times for raising will lower fabric strength. Nevertheless, if the raw material is inferior and a shorter staple length is used, then raising times may have to be increased to obtain the correct finish. Abrasion tests may be misleading on raised fabrics since the surface pile is rapidly worn off in the test before fabric rupture occurs.

REFERENCES

1. Thompson, 'Dyehouse Laboratory Practice' (Bradford: SDC, 1983).
2. British Standards Handbook No. 11.
3. McDonald, J.S.D.C., **96** (1980) 486.
4. Park and Thompson, J.S.D.C., **97** (1981) 523.

CHAPTER 12

Conclusion

As regards cost, fabric dyeing is intermediate between stock dyeing and yarn dyeing. However, an economic review is perhaps misleading for wool fabrics since several specialist finishing operations are involved, without which the full range of fabric qualities cannot be produced. It is difficult to compare the cost of fabric dyeing and finishing with dyeing at earlier stages. As has been indicated in earlier chapters, continuous processing can lead to economies in production. It is thus likely that future development will be concerned with the establishment of continuous processing lines and, in common with other sections of the dyeing industry, more use will be made of machine control and automation. This is not necessary solely to reduce labour costs, but also to overcome the shortage of skilled operatives.

Recent years has seen a move away from specialist production into the realms of mass production. At the same time there has been a reduction in the number of wool processing units, and those that remain are smaller so that capacity has decreased in real terms. This can be offset by increased production methods. In the last 30 years, in a vertical manufacturing operation from fibre to finished fabric, production per operative has been doubled or trebled in real terms by the introduction of modern machinery, automation and modern working practices. Whereas as late as the 1970s it was possible to find woollen manufacturers producing 2 000–3 000 pieces (150–240 linear km) per week, such producers have largely disappeared and only smaller units remain. However, in some parts of the world, where large-scale production of a small number of standard fabrics is more common, the adoption of continuous processing techniques can be expected.

Modern versions of traditional machines are now available. Thus scouring, milling, drying and finishing equipment is now constructed in modern materials using the best features of modern design and engineering practices. For example, the traditional winch can be replaced by the jet machine with advantage because of savings in several areas, of which resources is but one. Continuous processing ranges for carbonising and other wet and dry processes are available.

Failure to install or re-equip with modern plant has largely been due to the uncertain financial climate. However, if feasibility studies are undertaken, it is likely that the smaller unit that is now a feature of this section of the industry could realistically contemplate re-equipping and obtaining short pay-back periods.

It is hoped that this monograph has illustrated that the processing of wool fabrics is based on sound technology and is no longer a craft-based hit or miss operation. The industry is now firmly based in science and is supported by sound technology introduced to the industry by technical centres such as those of IWS. This technology is applicable on a wide range of modern equipment, so that techniques such as blind dyeing, the use of instrumentation and automation can be adopted for the production of wool fabrics to high standards and specifications, in competition with fabrics based on other fibres.

The future of wool fabric production is assured, and hopefully the modern technology and plant that is available will be exploited by the far-seeing industrialist and manager.

Index